Insects, Hygiene and History

Insects, Hygiene and History

by

J. R. BUSVINE

Professor of Entomology, London School of Hygiene and Tropical Medicine

THE ATHLONE PRESS of the University of London
1976

Published by
THE ATHLONE PRESS
UNIVERSITY OF LONDON
4 *Gower Street London* WC 1

*Distributed by Tiptree Book Services Ltd
Tiptree, Essex*

*U.S.A. and Canada
Humanities Press Inc.
New Jersey*

ISBN 0 485 11160 8

*Text set in 11/12 pt Photon Times,
printed by photolithography, and bound
in Great Britain at The Pitman Press, Bath*

Preface

Personally, I have enjoyed compiling this somewhat unusual book, over a period of some two decades. It has therefore been disconcerting when potential publishers or candid friends enquired: 'Fascinating, no doubt, but what sort of person would actually want to read it?' Perhaps a similar question was put to the late Hans Zinsser when he completed his entirely successful *Rats, Lice and History* and I wonder how he replied. Inspection of library shelves reveals at least some unorthodox subjects. For example, I know of two reasonably scholarly books which deal with the history of the W.C.; not, one would suppose, a prepossessing literary topic. No doubt the final criterion is readability, and as to that I cannot say. But should anyone wish to gain a vague idea of the subject, he will probably flip through the pages to find the illustrations. A sounder way might be to read the Introduction, which explains how I myself became fascinated with the subject.

Finally, a few words about the title. 'Insects, Hygiene and History' has the advantage of brevity and combines a fleeting reference to an earlier work of mine, with homage (by imitation) to Hans Zinsser. Since, however, the book deals with itch mites, as well as bugs, lice and fleas, it is not strictly accurate. The word 'vermin' is likewise incorrect, though I have occasionally used it where the context calls for a derogatory term; elsewhere in the book, I have used the rather academic word 'ectoparasites'.

J.R.B.

Acknowledgements

Professor P. C. C. Garnham and Dr. Robert Traub have kindly read and provided helpful comments on certain earlier chapters. In compiling material for the later chapters, I feel obliged to acknowledge the help of earlier writers who have summarised particular aspects of my subject. They include Bodenheimer, Cowan, Feytaud, Friedman, Hirst, Lehane, Madel, Oudeman, Raven, Singer, Weidner and Zinsser. If the great Montaigne compared himself to the string which bound up a bunch of other men's flowers, I can scarcely claim credit for much more than selection and arrangement.

More specifically, I found it necessary to visit perhaps a dozen London libraries and made many demands on the librarians. Most of them were unfailingly helpful and I would particularly like to mention those of the Wellcome Institute for the History of Medicine and the Institute of Classical Studies.

I am indebted to the Wellcome Trustees for permission to reproduce many of the illustrations (acknowledged in their captions) and to the British Library Board for Figure 7.3. Other photographic illustrations were prepared from various sources by the Visual Aids Department of the London School of Hygiene and Tropical Medicine. The various line-block figures were re-drawn by myself from various acknowledged sources.

Contents

Introduction

The topics dealt with in this book lie round the fringes of entomology, a subject which an elderly friend of my mother persistently confused with egyptology and which even editors, in moments of aberration, confuse with etymology. Such mistakes occur because entomology is seldom considered, except as an egregarious hobby, for many years personified by a bearded gentleman pursuing butterflies. Such an occupation has always appeared rather cranky in England. Thus, Moses Harris in *The Aurelian* (1768), discussing the butterfly *Glanville fritillary*, remarks: 'This fly took its name from the ingenious Lady Glanville, whose memory had like to suffer from her curiosity. Some relations that were disappointed by her will, attempted to get it set aside by acts of lunacy; for they suggested that none but those who were deprived of their senses would go in pursuit of butterflies. Her relations and legatees cited Sir Hans Sloane and Mr Ray: the last gentleman went to Exeter and at the trial satisfied the judge and jury of the lady's laudable enquiry into the Wonderful works of the Creation, and established her will'. This was about 1700. A century later the *Introduction to Entomology* by William Kirby, Rector of Barham and William Spence, begins rather diffidently: 'One principal cause of the little attention paid to Entomology in this country has doubtless been the ridicule so often thrown upon the subject ... in the minds of most men, the learned as well as the vulgar, the idea of the trifling nature of his pursuit is so strongly associated with the diminutive size of its object, that an *entomologist* is synonymous with everything that is futile and childish.'

Amateur entomology has, nevertheless, a long and distinguished history; witness for example, the 140-year-old Royal Entomological Society of London. Most of its devotees have concerned themselves (especially in former times) with the more aesthetically presentable insects, notably of course the glamorous lepidoptera. The majority of professional entomologists, on the other hand, are constrained to study species which are pests of one kind or another. These come in various shapes and sizes and some of them could perhaps be regarded as rather unpleasant. Among them are the subjects of this book: bed bugs, fleas, lice and itch mites, which have been among the main interests of my professional life. Most people consider this very queer indeed and regard with amused tolerance a fellow who can spend his time fiddling with such unpleasant creatures and who may, indeed, be wearing gauze-topped boxes of them under his socks, to supply them with meals of his blood. As for myself, I approached this queer profession by such gradual stages that it no longer seems odd. The first steps were not begun with a special preoccupation with insects, but rather with an interest in biology in general. This almost certainly started with the publication of the *Science of Life* in about 50 fortnightly parts, about 1929. The editors, H. G. Wells, G. P. Wells and Julian Huxley, made an excellent job of conveying the fascination of their subject; in particular, the story of animal evolution. As a result, I began the Special Degree course in biology at the Royal College of Science in London, where one of my heroes had already studied, as well as the grandfather of another. My specialization in Entomology, on the other hand, was largely dictated by economic necessity. Grants for students were non-existent in those days, except for a few scholarships, and the whole of my university career was financed by my widowed mother (except for a loan from the College, subsequently repaid). It was therefore necessary to get employment as soon as possible after graduation; and jobs were scarce in 1933 even with a first-class degree. After some nine months I obtained a research grant provided by Imperial Chemical Industries, whose staff I eventually joined. This began my prolonged interest in insecticides and their mode of action.

The next chance occurrence, which affected my career in a great and very fortunate way, arose from the fact that in 1939, the late Professor P. A. Buxton saw a need to bring up to date investigations on control of human lice, which have always tended to become serious and dangerous pests in time of war. In casting about for an assistant to test some new ideas in use of insecticides he came upon me. This began my career in medical entomology and my long association with Buxton, whom I revered and from whom I learnt much from example and precept. Now, nearly 40 years after beginning an interest in entomology, I have achieved a modest renown, albeit in a very narrow field. It is part of the purpose of this book to give thanks for the happy accidents that led me to this work, by attempting to show that specialization need not blind one to wider interests connected with one's particular studies. This has been a tenet of mine for many years. In 1943. I began an article in *Nature* as follows:

The amusing definition of a specialist is 'one who learns more and more about less and less' does not hold for the specialist in applied science. On the contrary, so far as I have observed in applied biology, it is a curious paradox that concentration on a single problem often requires the widest superficial knowledge of many subjects. For example, for the past three years, I have been studying the effect of new insecticides on the louse. That would appear to be a fairly specialized problem requiring merely a familiarity with the biology of the louse and the physical and chemical properties of the organic chemicals proposed as insecticides. Actually the research led to a consideration of such diverse subjects as the psychology of the louse (and, to some extent, that of the infested men who came to us for treatment), the risks of dermatitis from various chemicals and the possible carcinogenicity of oil diluents. The powers of absorption and retention of cotton wool and cellulose acetate underwear were considered and practical problems of laundering and dry-cleaning (especially the preparation of emulsions and the use of large centrifuge) came into the picture. Finally, one had to learn a little of such diverse subjects as army hygiene and women's hairdressing styles.

It is not the intention of this book, however, to deal with such ancillary problems, bizarre though they may have been, because all of them tend to be utilitarian; whereas it is my intention to dwell on curiosities of the subject, which have struck me as interesting, though of no discernible practical importance. It should not be necessary to defend what Sir Ray Lancaster used to call 'armchair science', for it is a wider background of knowledge which distinguishes the scholar from the artisan or, indeed, the scientist from the technician. On the other hand, it seems likely to be considerably harder to write a worthwhile book, which is at once scholarly and readable, than to produce a useful textbook.

The topics which seem worthy of animadversion, fall into two categories, roughly corresponding to C. P. Snow's 'Two Cultures'; that is, there are literary and historical aspects on the one hand and those of biological science on the other. Of most immediate appeal, no doubt, are the curious human reactions to these intimate little pests, throughout the ages. Just as archaeologists can reconstruct a bygone culture from surviving fragments of trivia, it is possible to learn much from human attitudes to insect vermin. Hints of their prevalence will illuminate aspects of social habits, and of modes and manners. They will provide literary curiosities as symbols of piety, of love, of human insignificance. They are subjects of ribald verse, of quack medicine and of morbid imagination. Again, the extent to which men's objective curiosity was focussed on these trifles, in different epochs, provides a continuous commentary on the long history of biological science, from Aristotle to the present day. Finally, and only during the last century, did it dawn on an incredulous medical faculty that these objects beneath contempt could transmit two of the most appalling epidemic diseases: plague and typhus.

Insects as vectors of disease turns our attention to their biological peculiarities. We can speculate on the circumstances in which these anomalous species became parasitic on man in prehistoric times, or indeed, took to a parasitic way of life, still earlier. Such considerations involve an outline of the family tree of insects and mites, since the creatures concerned are merely leaves on this enormous tree which, though on different branches, happen to converge in their close association with man. Even a brief outline of the evolution of arthropods (of which insects are the outstandingly successful and prolific class) implies some indication of their present numbers and variety. These remarks must be brief and statistical, though it is tempting to digress into their great range of queer shapes, habits and habitats. This almost infinite variety is all the more surprising in that, in many respects, insects and mites share our common biological requirements in respiration, digestion, excretion, nervous activity and reproduction; and there is surprising similarity in the biochemical processes involved in these functions. Nevertheless, it appears that the insect's experience of its environment and its reactions to it are so different from ours, that an astronaut's sensations when walking on the moon would seem relatively commonplace. The reason for this is the small size of all insects and I have felt that an explanation of the reasons for that (and its consequences) deserve some discussion.

Perhaps these biological sections may seem digressions from the main theme of the book. Yet something must be said about the whole kingdom of insects and mites to form a background to the account of their somewhat degenerate members. In the same way, an account of pimps and prostitutes of ancient Rome should begin with some remarks on the Roman civilization of the time.

1 The World of Insects and Mites

Throughout this chapter, I shall be continually referring to the fact that insects and mites live in a smaller size range than ourselves. This has a number of interesting consequences, but the first and somewhat ironic result is that most people are sublimely unaware that insects are our most successful rivals on this planet. Insects appeared here several hundred million years before us and there are distinct possibilities that they will survive long after we have starved, bombed or polluted ourselves into extinction. Yet because they are so small, it is difficult to recognise them as highly successful and relatively advanced forms of life. This claim is not based on their vast numbers, a criterion which would favour relatively limited organisms such as bacteria; it is their innumerable and complex ways of exploiting various environments which is impressive. Human intelligence has achieved this by a flash-in-the-pan brilliance over a very few thousands of years and the almost too successful dominance we have achieved is, perhaps, of a precarious nature. Insects, on the other hand, have percolated into almost every conceivable way of life and become firmly established therein, by the slow but remorseless trial and error of evolution. The result is an almost unimaginable variety of structure and habits in different species.

The numbers of species of insects which have been described so far is generally estimated at about 750 000; this compares with a mere 6300 reptiles, 8600 birds and 3700–4500 mammals. Indeed, the number of new species of insects described each year is in the region of 7000, which is comparable to the total numbers of the other groups mentioned. Most of these figures come from the 1972 edition of the Biology Data Book,[9] to which non-biological readers will perhaps accord an unrealizable infallibility, just as I cannot question a statement that an electron can hop from one orbit to another, spending one hundred millionth of a second in each; or that a quasar may be 1500 million light years away. A little thought, however, will reveal the difficulties of accurately enumerating animal species. To begin with, it has taken biologists a long time to reach a consensus of opinion on what exactly constitutes a species; and even now that there is some agreement on the level of genetical isolation which divides a race or variety from a 'true' species, there is rarely sufficient information to make such decisions. Another difficulty is continual growth of numbers of species due to the activity of naturalists and systematists. Thus, almost as many species of *Drosophila* fruit flies were discovered in the eight years preceding 1969 as in the previous 163 years;[204] many of these were named by a group of entomologists in the Hawaian islands, which are now estimated to contain no less than a quarter of the 1254 known species.[115] Similarly with larger groups, the Biology Data Book for 1956 quotes 15 000 species of *Arachnida* (spiders, mites, scorpions, etc.), while the 1972 edition gives a figure of 57 000. Yet this is far exceeded if we add together recent estimates of two major groups: 30 000 mites[174] and 50 000 spiders.[239]

Despite all these qualifications, the outstanding variety of insects is quite unchallenged; they are the most varied form of life known. Enormous numbers of species are found within the different orders of the class, especially the more advanced and successful types. Dr Paul Freeman, Keeper of Entomology at the British Museum (Natural History) has kindly assembled the following list of recent (1974) estimates:

	No. of species
Orthopteroid orders	
(Grasshoppers, locusts, cockroaches, etc.)	30 000
Phthiraptera (*Mallophaga*, biting lice)	3000
(*Anoplura*, sucking lice)	250
Hemiptera	
(Bugs, aphids, cicadas, scale insects, etc.)	61 000
Coleoptera (Beetles)	360 000
Diptera (Flies, midges, mosquitoes)	90 000
Siphonaptera (Fleas)	2000
Lepidoptera (Moths, butterflies)	160 000
Hymenoptera (Bees, ants, wasps, etc.)	150 000

Of these vast hordes of insects, only a minute fraction come to the attention of anyone but an assiduous entomologist. We notice the showy creatures like butterflies, the domestic forms like the housefly, and various pests. It is true that insect pests do a great deal of harm to agricultural crops, to forests and to food stores; and many are important in public health as disease vectors. But the actual number of species involved is comparatively small. A recent estimate of the more important pests gives a figure of 4500,[59] which is only a tiny proportion (about 0.6%) of known species.

The rather specialized and degenerate insects and mites which form the subject of this book do not belong to the groups which are very numerous and varied. It is true that the bed bug belongs to the large order *Hemiptera,* which includes a great variety of free living species, nearly all of which feed on plants. But the particular blood-sucking family concerned is small and rather isolated, containing about 75 species;[308] while the human flea belongs to a family with 122 species within a small and remote order of about 1800 species.[147] The family of human lice is very small indeed with only two or three species in two genera; these in turn belong to the quite small order of sucking lice.[88] Finally, there is the itch mite assigned to a family with 28 species,[258] and even this number would be reduced by some authorities who consider close relatives of this mite, which attack various domestic animals, to be varieties rather than separate species.

Abundance

The most impressive example of insect abundance is probably to be found in the huge swarms of locusts. A single, medium-size swarm of the desert locust (*Schistocerca gregaria)* will consist of some thousand million insects; and it is estimated that they can consume 3000 tons of food daily. Other examples[39] of insect abundance include the Hessian fly (*Phytophaga destructor*), certain harlequin flies *(Chironomus plumosus)* and biting midges *(Culicoides dovei),* all of which may occur in numbers ranging from 3 to 5 million per acre. Even larger numbers of tiny insects (mainly primitive 'springtails') occur in soil; populations estimated per acre of arable soil in England reaching 7.7 million and in Denmark up to 34 million, with an average of 15 million. Local aggregations of insects include: 30 000 or 40 000 honey bees, ant nests with over 100 000 inhabitants and termite mounds with up to 1.7 million individuals. Even in the upper air, collections of drifting specimens of diverse insects indicate populations of some 25 million per square mile. As regards mites, there are comparatively few estimates; but where they occur, their numbers are enormous. In soil, for example, they greatly outnumber all other arthropods (70–80% of the total), and their actual weight may constitute about 7% of the invertebrate fauna.[82]

Normally, populations of parasitic insects are comparatively small. I do not know of any estimates of infestations of bed bugs or fleas; but my impression of those I have seen relates to hundreds rather than thousands. There are, indeed, records of exceptionally heavy infestations of body lice; (3800, 10 428 and even 16 822) but these are very exceptional indeed. Surveys recorded by my former chief, the late Professor Buxton,[54] show that infestations over 100 are quite rare and most infested people carry between 10 and 20 lice. Similar findings relate to head lice; maxima of 1286 and 1434 have been recorded, but most cases had fewer than 10 per head. The numbers of itch mites on scabetic patients can only be estimated from the adult female mites. Among 900 adult male patients the average infestation was only 11.3, with only 3% carrying over 50 mites.[212] It seems that populations of ectoparasites are not limited by competition for food, or the attacks of other parasitic or predatory insects, such as check the proliferation of most free living insects and mites. Insect vermin tend to be kept within bounds by the cleansing activities of their irritated human host. This may take the form of combing out head lice, while body lice may be sought along the seams of garments and cracked between the finger nails, or even teeth, of infested people. Scabies mites, on the other hand, tend to be dug out and killed by the scratching finger nails of sensitive sufferers. In an analogous way, the less closely parasitic bed bugs and fleas tend to be decimated by ordinary household cleansing. Both bugs and fleas have become rarer in civilized countries with improved housing and cleansing facilities, such as the vacuum cleaner. Body lice have disappeared from the great bulk of citizens who regularly wash and iron their underwear; but head lice can persist even with regular washing, since the temperature used is too low to kill the insects.

Size

The largest insects reach a length of 120–150 mm, about the size of a mouse and certainly larger than the smallest vertebrates (small frogs, of the South African genus *Phrynobatrachus* are less than 25 mm long). Among the largest insects are the rhinoceros beetle *Dynastes hercules* (155 mm), a stick insect *Pharnacea serratipes* (260 mm), a bug *Belostoma grande* (115 mm) and a moth *Erebus agrippina* (wing span, 280 mm).[263] The smallest insects include certain beetles of the family Ptiliidae and 'fairy flies' of the family Mymaridae (which parasitise eggs of larger insects). One species *Alaptus magnanimus* measures only 0.21 mm in length. As Julian Huxley[154] pointed out, these tiny parasitic wasps are 'of smaller bulk than the human ovum and yet with compound eyes, a nice nervous system, three pairs of jaws and three pairs of legs, veined wings, striped muscles and the rest!'

The great majority of acarines are mites which, as already remarked, have exploited a smaller size range than the insects. Most of them fall in the size range, 0.2 up to 2 mm in length but a few exceed this and, in particular, the families of ticks include large forms (e.g. *Amblyomma clypeolatum* or *Ornithodorus acinus,* each being about 30 mm). Among the smallest mites is the male *Acaraspis woodi,* which is only 0.1 mm long.[150]

If we compare the insect vermin with these sizes, we find the bed bug to be about 6 mm long, the human flea 4 mm (other fleas, 1–5 mm), the human body louse about 4 mm (head louse, 3 mm), while the itch mite, adult egg-bearing female, is only 0.4 mm. The weights of unfed adult females of these various creatures would be approximately as follows: bed bugs, 3.5 mg and head lice, 1 mg; actual records of these have been made. The tropical plague flea weighs about 0.6 mg so that the human flea, which is larger, must be about as heavy as a head louse. As regards the itch mite, I can only calculate, on the basis of its diameter and assuming a specific gravity of one, that it must weigh about 0.01 mg.

Size and the Environment

It has been remarked that the size range of insects in the animal kingdom, stretches from that of the smallest vertebrates down to the largest protozoa.[91] The smallness of insects and mites has an interesting cause and some remarkable consequences, which deserve closer attention. To begin with, their tiny size offers them a much larger number of different habitats in the world, thus providing opportunities for their great variability. Thus, a small farm can be regarded as a characteristic basic habitat for a man or, indeed, for his domestic animals. But within such a farm, there will be dozens of distinct biological niches for insects and mites. Large numbers live in the soil, either burrowing into plant roots, feeding on organic debris or predatory on other forms. Many attack vegetation above ground; larvae feed on leaves or burrow into stems or fruit (some indeed burrow a minute tunnel in the thickness of a leaf). Other forms live under tree bark or burrow into the wood and some stimulate the plant tissues to produce galls in which they live and feed. Large animals are another source of food and shelter; parasitic insects and mites live among the fur and feathers of beast and bird or even burrow into the skin or deeper tissues. Many forms, however, merely breed in the dung.

The fact that insects are so small and that many live in hidden niches explains why the vast majority escape our notice, which is even more true of the still tinier mites. Add to this the fact that men congregate in cities and know little of the vast tropical forests and bushland teeming with insects and we get the impression of two distinct forms of life co-existing but almost in different dimensions. Only one or two in a thousand affect us or our crops sufficiently to be called a pest; and about one in ten of these pests cause serious depredations or carry dangerous diseases. True, the problems raised by this few hundred major pests are grave enough; but it took mankind a long time to recognise the fact. The depredations of a swarm of locusts are obvious enough, but it is less than a century since we began to realize the menace of the lowly louse, the contemptible flea or the fragile mosquito. It is not only difficult to recognize their danger as disease vectors, it is hard to accord them the dignity of organized animals and to realize that when we 'swat' a fly we obliterate an organism at least as complex as a modern computer.

Reasons for small size and its consequences. The reasons why insects and mites are all so small are ultimately dependent on the adoption of an external skeleton by the arthropods and of an internal skeleton by vertebrates, a dichotomy which occurred an immensely long time ago. In this remote period, two groups of animals discovered an extraordinarily effective device for assisting locomotion; the skeleton, which is a framework of jointed levers. The arthropods formed theirs from the integument, whereas the vertebrates developed internal cartilaginous rods which eventually became bones. In some ways, the external skeleton had advantages: not only did it provide more area for muscle attachments, but, like a suit of armour, it provided some degree of protection. But an external skeleton has one important drawback; the rigid portions cannot expand to allow for growth. The arthropods solved this problem by moulting at intervals, inflating themselves with water (later, in terrestrial forms, with air) and forming new, expanded hard integuments. Before the new cuticle hardened they were virtually without a skeleton; but this is much more serious for a large animal than a small one, for dimensional reasons illustrated in the following table.

Relative dimensions of humans, insects and mites

Dimension	Mite	Insect	Human	Ratios Human/insect	Human/mite
Length	0.75 mm	7.5 mm	1500 mm (59 in)	× 200	× 2000
Width and Depth	0.1 mm	1 mm	200 mm (8 in)	× 200	× 2000
Area	0.01 mm²	1 mg	40 000 mm² (62 in²)	× 40 000	× 4 million
Volume	0.0075 mg	7.5 mg	60 kg (130 lb)	× 8 million	× 8000 million

This shows that, if we compare a human being with an insect such as a bug or louse, which is of the order of 1/200th in linear dimensions, the human is 40 000 times as strong (cross-sections of muscles or bones). But, and this is the crucial point, the human is 8 million times as bulky and heavy. Hence, we should find much more difficulty in moving our bodies about if we were temporarily deprived of our skeleton than would a very small animal. The helpless condition of moulting arthropods, depending on the rigidity of living tissues, therefore explains their size limitations, the largest land forms being only a few centimetres long. (Marine arthropods can grow larger, since their bodies are supported by the weight of water displaced; and some spider crabs can grow to about 50 cm). The size limits imposed on land arthropods has caused insects to exploit different living conditions; and the mites, a dimension smaller, have chosen habitats still more remote from our own. Certain interesting consequences follow from these changes in scale.

Apparent strength. From time to time, someone points out the wonderful strength of ants or 'performing' fleas, which can carry loads many times their own body weight, or jump many times the length of their bodies. It is, however, evident that this is merely a dimensional effect as explained in the previous paragraph. In fact, the muscular power of such insects as have been measured are comparable to those of vertebrates, when calculated as pull per square centimetre.

Another less obvious result of the apparent strength of small creatures is their relative immunity to injury. During the first World War, a German entomologist who was studying the human louse with painstaking teutonic thoroughness, discovered that to squash an unfed female required a weight of some 1.5 kg, or about a million times their own weight!

Optimum temperatures. Small creatures like insects have a large surface-to-volume ratio, so that if the human body in the above table was turned into an equivalent weight of insects, these would require 40 000 times as much skin; and this has several consequences. Thus, an insect has relatively enormous radiating surface, and it would be physiologically impossible to maintain a constant body temperature significantly above that of the environment, as the higher vertebrates managed to do; which means that their biological processes are much more severely regulated by the ambient temperature. In simple terms, when it is cold, they are torpid; but as the temperature rises, they live faster and faster (and usually more efficiently) up to optimum, above which the overloaded metabolism becomes impaired. There is usually a relatively limited temperature range (say, about 20°C wide) over which insects and mites can live more or less normally; outside this range the temperature is potentially harmful, though short exposures can be endured. The position of the normal range varies, being lower for insects in cool climates and higher in tropical ones. The human bed bug is adapted to a warm climate with a range of about 15–30°C. The human flea is difficult to rear in captivity and its requirements have not been so carefully studied, but from some early studies of A. Bacot, its extreme lower limit seems to be 10°C, compared with 4°C for the common rat flea and 16°C for the tropical plague flea.[17] Human lice and itch mites are quite abnormal in their temperature requirements. A long adaptation to living on (or in) the skin has conditioned their preference to 30°C and they can only thrive at temperatures close to this.

Cool conditions. The temperature which is favourable for normal growth and development is not the same as that which allows maximum survival of these parasites during starvation. In most cases, this is about 13°C; lower temperatures are themselves harmful, and at higher ones, starvation takes effect more quickly. There are, nevertheless, big differences in this maximum survival at 13°C. Lice and itch mites live on their hosts and can feed at any time and these 'pampered' parasites can only live for about 14 days without food; but human fleas have survived starvation for 105 days[17] and bed bugs as long as 360 days under cool conditions.[160]

Extreme cold. Mammalian tissues are not especially sensitive to low temperature, since blood or semen can be stored without harm at $-80°$ to $-190°$C; but the lower limit for whole animals, as judged from experiments with hamsters, is about $-5°$C. In very cold climates, the body temperature is maintained by metabolism, protected by insulation (hair, fat layers).[87]

Insects, too, avoid very great extremes so far as possible, by hiding in microclimates (even protection by layers of snow). But even so, they may have to withstand temperatures down to $-40°$C if they live in arctic or mountainous regions. Hibernating insects protect themselves by producing glycerol in their tissues, before winter. This lowers the freezing point, as it does in a motor car radiator; but this is only a minor part of the protective mechanism, which is not fully understood. Limited short exposures to cold can be endured by insects normally adapted to a warm environment. Thus, the bed bug can endure exposures of nearly 2 hrs at $-17°$C. Human lice require $7\frac{1}{2}$ hours at $-10°$C and 2 hrs at $-15°$C to kill them. The egg stage is more tolerant and needs 10 hrs at $-15°$C.[52]

Extreme heat. Large animals in hot countries take refuge in shade, of course; but they are generally able to cool themselves by perspiration. This is not feasible for insects, because they would soon suffer from desiccation if they used up their small volume of water for cooling. They too seek shade and partly insulated micro-climates, but some exceptional insects have to endure great heat. Thus, in deserts, mantids and grasshoppers can walk about on soil at $50°$C[53] and larvae of certain chironomids or harlequin flies have been found breeding in hot springs at this temperature.[38] The most extreme case of insect tolerance of heat, however, is a midge which breeds in rock pools in central Africa. When these dry up, the larvae may have to endure a day temperature of perhaps $60°$C. It does this by turning into a dried-up inanimate state, in which it can survive for as much as 39 months, reviving soon after being put back into water. In their desiccated condition, they can survive 5 minutes exposure to $100°$C.[134]

The insect vermin which are my principal concern are not adapted to extremes of heat. The temperature becomes harmful about $40°$C and lethal at $45°$C. Half an hour at this temperature will kill bugs, lice, fleas and itch mites, as well as their eggs.[52]

Water relations. Owing, once again, to their very high surface: volume ratio, insects and mites would rapidly perish from desiccation if their vital body fluids were not protected from evaporation. This is achieved by the special properties of their integument, the outer layer of which consists in a thin (0.5μm) layer of highly stabilized protein, lipid and phenolic substances. This carries a layer of wax and above this (and penetrated by it) a kind of varnish or cement. The wax consists of long chain-like molecules, standing at right-angles to the surface and packed tightly together to form a waterproof layer. The integrity of this layer is essential for preventing water loss. If this film is disrupted by mechanical scratchings or by adsorption into micro-pores of certain silica powders, the insect loses water rapidly and may die; though if the damage is not severe, more wax may be secreted and the layer repaired.[320]

Respiration. A further consequence of the large surface-to-volume ratio is, however, a possible benefit. Since all parts of the body are near to the surface, it is possible to utilize respiratory systems depending simply on diffusion. Insects and some mites employ a system of small tubes, ramifying through the tissues and connected to the air by small holes, which can be opened or closed by muscular action. This is mainly for access of oxygen, since carbon dioxide in many small insects simply diffuses from the cuticle. Indeed many families of mites (including the itch mite) do not even need the ramifying tubes. In either case, the system is much simpler than the complex blood transport arrangement of vertebrates and eliminates the need for vulnerable heart, blood vessels and lungs. Furthermore, it dispenses with the elaborate nervous control of respiration which is so sensitive to deprivation of oxygen. Consequently, insects and mites are much more resistant to asphyxiation. Many years ago, I investigated this power in several insect species by keeping them in pure carbon dioxide or nitrogen for various periods. The bed bug, which was the most susceptible of the species tested, revived after an average of 15 hours, while the most resistant insects (which were grain weevils) survived 215 hours.[46]

Aerial transport. Unless we are encountering a strong gale or indulging in free-fall parachute jumps, we are relatively unconscious of the viscosity of the air. But this is quite important to small creatures with a large area : volume ratio. Objects falling through a gas reach a terminal velocity, inversely dependent, according to Stoke's Law, on the square of the radius. For small objects this terminal velocity is not excessive and insects are most unlikely to be killed by a fall. On the other hand, accidental displacement by wind is a distinct possibility, and collections from aircraft or kites reveal many small insects drifting in the upper air. Some species try to avoid this hazard by refusing to fly except in calm conditions; though wind may actually assist dispersal beneficially.

Size relations are important in relation to insect flight.[253] Four entirely different groups of animals have developed the power of winged flight; in historical sequence, insects, flying reptiles, birds and bats. Those insects which commonly use flight for locomotion, fly at speeds between 1 and 30 m.p.h. (Birds range mainly between 10 and 30 m.p.h., with a maximum about 60 m.p.h.). In both birds and flying insects, the flight muscles are substantially developed; on the other hand, they tend to atrophy in flightless forms, which do not need to expend food reserves on this part of the body.

Insects, like other flying animals, exploit the lifting power of a surface moved forward in air (aerofoil effect), as man was later to do in aircraft. This dynamic convenience is curtailed by two size limits. At the upper end, the weight increases faster than the power available from the muscular effort necessary to raise it, setting a maximum size about 25 kg. This is well above the size range of insects, imposed for other reasons, as we have seen. But another limitation occurs in the lower size range, due to a factor known as the Reynolds number, which concerns the flow of fluids over surfaces. Some 90 years ago, Osborne Reynolds (working on liquids travelling through pipes) noticed that turbulent flow began when the relation (length × speed)/velocity fell below a critical level. Turbulence spoils the aerofoil effect and insects have a problem in maintaining a suitable Reynolds' number because their wings are so short. They compensate by speed of movement; thus the wings of efficient flyers like houseflies and bees beat 200–300/sec; the record is a midge, reaching about 1000/sec. Some excessively small insects abandon the aerofoil effect and utilize a different motion, more akin to swimming; and a few can actually use the wings for swimming in water as well as in the air. Instead of normal membraneous wings, they have rods fringed with long bristles.

The mechanics of the very rapid wing movements mentioned above are interesting. The frequency of contraction does not depend on repeated nerve stimulation, but on the inertia and elasticity of the load in a resonating system. The cuticular framework which contains hinges or pads of an excellent elastic material called resilin, is set into vibrational movement. Not only does the resonance of the system regulate frequency, but the elastic deformations alternately store and release energy.

Recent investigations[24] have shown that flea jumps depend on a very similar mechanism, together with a pad of resilin, almost in the position of the hind wing hinge. Energy is stored by folding up of the hind legs, which compress their respective wing pads and reach a position where the power is locked, rather like a bent cross-bow. On movement of a trigger mechanism, the pent energy is released in a violent extension of the leg, much faster than would be possible by any muscular contraction. As a result, the flea shoots off at an acceleration of some 200 gravity, a process which has been studied by high speed cinematography. Despite all this efficiency, the jump of a flea seldom exceeds a height of 5 or 6 inches (maximum about 20 cm). Despite a rapid initial velocity (about 2 m/sec or 5 m.p.h.) the flea's small momentum is soon slowed down by its relatively large air resistance which subject begins this section!

Terrestrial movement. A few lines of doggerel, familiar to young entomologists, run as follows:

> The butterfly has wings of gold
> The fire-fly of flame.
> The little bug has none at all
> But gets there just the same.

The German entomologist, Hase, to whom I referred earlier, actually timed bed bugs running on a horizontal surface. He found that newly hatched nymphs covered 20 cm (8 in) per minute and adult bugs could move five times as fast. In relation to the length of the bug, this appears quite fast, and it is surprisingly difficult to pick them up with fine forceps. But translated to our own scale, these speeds correspond to only 0.01 and 0.05 m.p.h.

The other parasites with which we are concerned are even less speedy. Fleas cannot walk at all well on flat surfaces, though they scurry very efficiently through fur, feathers or clothing. On an inhospitable open space, they commonly resort to jumping, which I discussed in the previous section. The fastest lice crawl at about 23 cm per minute (0.01 m.p.h.)[54] and itch mites cover only about 2.5 per minute (0.001 m.p.h.).[212]

It goes without saying that these modest walking speeds are perfectly adequate for their regular needs, in the search for food or mates. Lice and itch mites spend their lives very close to the host and merely need small trips round the body (in some experiments, about 20 years ago, I found that adult lice released at a certain point under my clothes had moved an average distance of about 25–30 cm two hours later).[47] Adult fleas spend some time on the body or among clothing, and the rest close to the host's sleeping place; usually among bedclothes. Bed bugs have the longest regular journeys to make; but experiments with artificial huts 'baited' with rabbits, as well as examinations of natural infestations, show that most bugs live in bed crevices or in the wall adjacent.

So much for *regular* movements; but these will not spread the parasites to 'fresh woods and pastures new'. For this purpose, they have to rely on passive transport. It is interesting to note that a few insects and a large number of different kinds of mite make frequent use of rides on larger arthropods, usually flying insects, a practice known as 'phoresy'. The parasitic forms, however, depend on their hosts. Lice and itch mites, which keep close to the body, reach new feeding grounds during close contact between their host and another human being (especially when they sleep together). Bugs and fleas rely on movements of the host's belongings, especially furniture or bedding. It was on bedding bundles that bed bugs invaded many air-raid shelters during the Second World War.

Water surface film. The small size of insects involves a large linear : volume ratio, which is important in relation to the surface tension of water. Like other large animals, we plunge into water without noticing any impediment to penetrating the surface film; but the mass of an insect being so small in relation to its outline area, it is quite difficult to push it under the surface. To emphasise this fact, I have encouraged students to measure surface tension by a method which involves pulling a fine wire ring through a water surface. A ring 2 cm in circumference (very roughly equivalent to an insect outline) requires a force of nearly 300 dynes (corresponding to a weight of about 300 mg) and I invite them to compare this with the weight of a mosquito or unfed bed bug (less than 5 mg) or even a housefly (about 15 mg). Small wonder that insects are seldom drowned; but if they *do* get wet, they have to struggle to escape!

Adhesion. Very small particles adhere to each other and to surfaces, due to van der Waals forces. Even greater adhesion is obtained with soft or elastic bodies, which increase the area of mutual contact. Furthermore, under humid conditions, the adhesive forces are reinforced by surface tension, since a thin film of water will be formed between the surfaces. Actual experiments with 100 μm diameter glass spheres, show that only 75% of them are detached by a force of 4.7 dynes (and scarcely any by 0.39 dynes).[67] Calculations show that for 100 μm spheres with a specific gravity of 1.0, the respective forces correspond to 8750 and 750g (or 8750 or 750 times their weights).

The sizes I have mentioned are very small compared with most insects; but they are not far below those of mites. An adult itch mite averages about 300 μm in diameter and of course eggs and larvae would be much smaller. It is therefore easy to see why an itch mite extracted from its burrow adheres to the needle; also, why skin-living mites do not need to cling tightly to prevent themselves falling off.

Perception of the Environment. So far, I have been discussing mechanical and physical relationships between insects and their environment; but it is also important to try to imagine something of the ways in which they perceive their surroundings. This is, of course, a very large subject to which books have been devoted, or at least, large sections of textbooks on insect physiology. In this account, I want merely to make a few general comparisons with our own experience of the world and those of insects or mites, especially parasitic vermin.

Firstly, these small creatures rely on similar senses to our own; sight, touch, smell, taste, hearing, perception of warmth and cold, though in most cases the quality of the sensation is likely to be distinctly different. Thus, touch usually involves distortion of bristles and must be something akin to our feelings when something touches our hair. Sight is dependent on a different optical system. Hearing is often absent and when it exists it seems to be crude.

Secondly, various senses are developed to very different levels in different insects and mites. In what we would regard as the most valuable sense, that of sight, even the most highly evolved adult insects cannot claim parity with us. Certain blood-feeding insects, such as mosquitoes or tsetse flies are relatively well endowed in this respect; but the parasitic insects and mites with which I am concerned are purblind or blind. At best, their vision approximates to what we see through a ground glass screen. Even the other senses of these creatures seem only to operate over a very short range. The sense of touch and avoidance of light are used to find a hiding place. Sensations of warmth and smell to detect a nearby host. It is therefore evident that we must discount the folklore stories of bugs systematically travelling long distances to seek a meal (e.g. by intentionally dropping on to beds from the ceiling). To a considerable extent, infestations of bugs, lice, fleas and itch mites are spread by movements and contacts of their hosts or their hosts' apparel or belongings. To this extent, they resemble parasitic micro-organisms, which are transmitted passively, by chance.

Intrinsic effects of small size. Two final consequences of small size are intrinsic, in the sense that they do not directly relate to the environment. The first is that, being small, insects grow to maturity quickly, the time required being anything from a few days to a few months (as compared with the several years required by vertebrates). This means that small steps in evolution can be more easily perceived. They are certainly convenient material for genetic studies.

The final consequence of small size of insects, is that their brains are small in relation to the size of a nerve cell (which is not markedly different in various classes of animal). As a result, it is very unlikely that an arthropod brain could ever achieve the inconceivable complexity of those of higher vertebrates. One advantage of the relatively simple nervous system, with its reduced dependence on one central brain, is that it is less vulnerable. Decapitated insects, though sluggish and sub-normal, live reasonably well until they die from starvation or desiccation. It is well to remember when we speak of a simple nervous system, that the brain of a small insect is able to perform tasks which one could reasonably compare, I suppose, with those of a modern electronic computer. A notable difference is that the insect brain may weigh only about a microgramme. Furthermore, though the innate chains of reflexes are not well adapted for modification by experience, they have been perfected by aeons of evolution to cope with most of the normal problems of survival. If we carefully observe the behaviour in some of the higher insects, it arouses certain uneasy feelings by its combination of efficiency and yet remoteness from our own mental processes. This is well expressed by a quotation from Maeterlinck cited by Brues.[39]

Insects do not belong to our world. Other animals, even plants, in spite of their silent life and the great secrets they sustain, do not seem total strangers to us. They are surprising, marvellous often, but they do not upset our ideas from top to bottom. The insect alone gives the sense of not belonging to the behaviour, the morals or the psychology of our world. One feels that they have come from another planet, more monstrous, more energetic, more ruthless, more atrocious, more infernal than our own.

2 The Origin and Evolution of Insects and Mites

So far, I have been considering the way of life of insects and mites as a group; but now I want to examine differences between the various forms and discuss the way in which this has come about in the course of evolution. Most people, of course, scarcely consider the matter; but if they do, they are inclined to lump together all ectoparasites as vermin. In the words of Dr Johnson 'there is no settling the point of precedency between a louse and a flea.' In an effort to correct this impression, I would like to show how different are their origins on the huge family tree of land arthropods. This is illustrated by the diagram in Fig. 2.1 in which the fleas, lice, bugs and mites can be found, with a little care, among the branches representing different natural orders. But first, something must be said about this evolutionary chart and the way in which it has been traced.

The Arthropod Family Tree

Elucidating the evolutionary history of an ancient group of animals like the arthropods is difficult and complex. There are two sorts of evidence, the fossil record and anatomical similarities. The fossil evidence is, of course, erratic; and, I suppose, the palaeontologist must be frustrated by the abundance of well-established forms compared with the rarity of important transitional 'missing links'. So far as comparative morphology is concerned, the main difficulty is to decide whether a similarity is evidence of a common ancestor, or of 'convergent evolution'; that is to say, the result of two unrelated forms independently adopting a similar structure for a particular biological need. Accordingly, it is not surprising that the experts do not always agree. In fact, the subject involves so many and complex different interpretations of the evidence, that it is only possible for me (with limited knowledge) to attempt a gross simplification, which I hope will not be too misleading.

It is generally agreed that the arthropods evolved from a group of

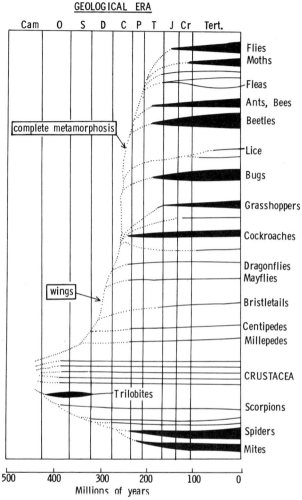

Fig. 2.1. Some of the main branches of the arthropod family tree. The firm lines show the geological period from the earliest fossils of various orders: dotted lines suggest putative relationships. The two main evolutionary explosions of the insects were associated with development of wings and complete metamorphosis.

Key to geological eras: Cam., Cambrian; O, Ordovician; S, Silurian; D, Devonian; C, Carboniferous; P, Permian; T, Triasic; J, Jurassic; Cr., Cretaceous; Tert., Tertiary.

segmented worms, which experimented with crawling instead of wriggling about, by growing a pair of legs on each segment. This different type of mobility put a premium on the development of a skeleton, or jointed rigid framework. The primordial arthropods formed this from a toughened outer skin, articulated at the joints like a suit of armour. But now a divergence of expert opinion arises, for some authorities[284] believe that the huge, diverse group of arthropods were descended from a single stock; while others consider the variation so great, that the arthropod armour plate innovation may have happened more than once, to give rise to different branches of the group.

It is very unlikely that the argument will ever be solved on the basis of fossil evidence, since there is virtually no relevant fossil record of these very remote events, which probably occurred in the pre-Cambrian, some 500 million years ago. Accordingly, the hypotheses are based on comparative morphology. For example, inferences can be drawn from different ways in which groups of segments have fused together to form specialized regions of the body. This nearly always occurred at the front end to form a head, where the segmented legs were changed to antennae, to jaws or to other feeding appendages; in addition, it often occurred in a subsequent section, where extra large legs took over most of the walking activities. Speculations on relationships between the different groups usually involve consideration of how many segments have been involved in these specialized clumps.

Figure 2.1 shows the evolutionary tree of some of the more important arthropods, based on some recent authorities. It will be seen that, while the arthropods originated in the sea, they have been extremely successful as terrestrial animals, especially in the two large groups, the insects and arachnids. The latter appear to have branched off from the ancient group of animals called trilobites vaguely similar in appearance to woodlice, which teemed in the shallow waters of the sea 400 million years ago, but which finally became extinct. The insects, on the other hand, appear to have come from a common ancestor with centipedes and millipedes and it is possible that their emergence to dry land was a separate occurrence.

The terrestrial environment produced several problems, the obvious ones being changed respiration and protection from desiccation. The respiration of the ancestral aquatic arthropods was probably managed by plate-like 'gills' attached to the legs. Something like this is retained in scorpions, spiders and a few minor orders; but the leaf-like gas-exchange surfaces are usually enclosed inside cavities, for protection, and are known as lung-books. Other land arthropods adopted a system of holes along the body connected to ramifying tubes designed to carry the oxygen directly to the tissues. This system apparently arose independently, in different groups, including a few terrestrial crustacea (certain woodlice) mites and the centipede-millipede-insect group.

The importance of preventing water loss in small terrestrial animals has already been mentioned and also the waterproofing wax layer used by insects and mites. This specialised cuticle shows some similarities in the two groups, which is probably an argument for their affinity.

With the primary adaptive problems solved, a vast and varied new environment was open for exploitation by the new land arthropods. Considerable advantage was obviously attached to improved locomotion and the outstandingly effective advance was the development of flight by the primitive insects. The mysterious origin of insect wings is still a subject for erudite speculation, such as that at a Royal Entomological Symposium in 1974. Insect fossils are little help since the earliest (325 million years old) have well-developed wings, somewhat resembling those of dragon flies or mayflies, and almost certainly fully functional. The most commonly accepted theory is that wings developed from flap-like projections on the upper edge of the second and third thoracic segments. This idea is supported by several early fossil insects with flaps on the *first* segment as well as normal wings on the next two. At the Symposium mentioned, however, Sir Vincent Wigglesworth rehabilitated a still older theory, dating from 1811, which derived wings from modified gill plates of insects with aquatic early stages.

However wings developed, it is likely that they began as fixed structures, which could have helped the primitive insects to parachute down from tall vegetation and land on their feet, so as to scurry away from danger. Wigglesworth, again, took a slightly different view, suggesting that gliding flaps were first exploited by small insects swept up into the air by wind. The main advantage of this would be in improved dispersion to new breeding sites. In either case, the later development of flapping movements would be likely to improve the performance and lead to true flight.

Whatever its origin, the power of flight was almost certainly responsible for the enormous success of the insect group. The great proliferation, both of numbers and varieties, was the first and perhaps major 'evolutionary explosion' in their history. Quite soon, however, a difficulty connected with moulting had to be overcome. Functional wings are large and fragile and present problems in moulting; indeed, only one existing group of insects (the mayflies) casts its skin after expanding the wings. Accordingly, evolutionary pressure soon induced insects to utilize the wings only in the adult stage, when they were most necessary for migration and colonization. This divided the life into a young growing stage without wings (except rudimentary pads, towards the end of growth) and a winged adult; the change to the 'perfect' insect is known as *metamorphosis*.

The first type of metamorphosis adopted by insects was not, however, radical. The young forms (conveniently described as 'nymphs') usually resemble their parents in many ways; they usually seek food in the same way and, for this purpose, have similar sense organs. But the next major evolutionary event took the insects further along the path of specialization and enabled them to benefit from a complete division of the life history into totally different phases. The young form (larva) was now adapted solely to feeding and growing and appeared quite different from the adult, which was more exclusively concerned with reproduction and (taking advantage of its wings) with migration and colonization. This is described as *complete metamorphosis;* and to permit the drastic remodelling of the body, a quiescent phase was introduced in the form of a pupa or chrysalis. In its extreme form, the change is impressive. The larva emerges from the egg as a maggot or grub, almost in the condition of a precocious embryo, blind and nearly helpless. It does not however need to seek its food, because the adult has deposited the egg in the midst of a plentiful food supply, just as the mother bird supplies food in an abnormally big egg and the mammal through the womb. After the quiescent pupal stage, the adult displays all the necessary organs of special sense and powers of accurate and sustained flight. It does not grow and it therefore does not need food to grow, though carbohydrate is usually required for activity and the females need protein for egg production. Sometimes no food is required, the reserves accumulated in the larval stage being sufficient. The relative success of these most highly evolved insects can be shown by the proportions of known species existing in the various evolutionary grades. Thus: wingless insects, 0.1%; primitive stiff-winged forms (dragonflies, mayflies), 0.9%; partially metamorphosing insects, with folding wings (cockroaches, crickets, bugs), 13%; completely metamorphosing insects (beetles, bees, flies, butterflies), 86%.[192] By contrast with the insect group, the non-flying land arthropods have not shown clear evidence of sharp and progressive evolutionary advances. The centipede and millipede groups descended from early ancestors of the insects, have changed little from their fossil progenitors; nor have they been very successful, as judged from numbers and variety of surviving types. The same can be said of the scorpions and certain minor groups of arachnids, such as the harvestmen and whip scorpions, little known except to experts. Two other groups of arachnid achieved modest success; the spiders and the mites and ticks.

Spiders solved the problem of long distance migration by drifting about on threads of their own silk, which bears the same relation to insect flight as ballooning does to transport by powered aircraft. The acari or mites adopted a considerably smaller size range than other land arthropods and movement to new feeding grounds presents a less urgent problem, since vast populations can build up in relatively small places. For transportation to new sites, they mainly rely on accidental carriage by other animals or dispersal by wind or water. In addition to adopting a small size range, the ancestors of mites abandoned the restriction to the purely carnivorous diet of other arachnids. An immense variety of different micro-habitats then became available to them.

As with insects, the harmful mites with which we may be more or less familiar, represent only a very small fraction of the 30 000 known species. It is quite impossible for me to give more than a rudimentary impression of the variety of forms and habits (some of them quite bizarre) of which most of us are much more ignorant than we are of the insect kingdom. An excellent introductory account is provided by the opening chapters of T. E. Hughes'[150] book *Mites or the Acari*. All that can be done here is to note some of the principal feeding habits, which will provide a basis for later consideration of the development of the parasitic way of life.

I suppose the great majority of mites are free-living inhabitants of soil or surface vegetation, feeding on plant or animal remains or on faeces. Many feed on fungi or bacteria associated with such organic debris. Then, there are forms that feed on living plant tissues (some forming galls in the process). Among these vegetarian and saprophagous forms are predatory forms attacking them and also other small creatures in the environment. These free-living mites extend through a great range of sites from nearly bare mountain tops, through forests, meadows, swamps and lakes to the shore and even the deep sea. Apart from free-living forms, there are many mites which enter into various types of relationships with other animals. The exploitation of insect mobility by 'phoretic' rides, has already been mentioned. Also, there are varying degrees of 'commensal' sharing of foods (passing into parasitism) of other arthropods. Larger animals are not neglected, partly for transport, but very often to provide food in the form of sebaceous secretions and other associations of hair or feather follicles. These have sometimes evolved into the harmful forms causing injury of skin surfaces, as in the mange mites and the itch mites.

Evolution of Mites

Since there is scarcely any fossil evidence of the evolutionary history of mites, we can only make speculations based on morphological resemblances used for classification. But the classification of mites is much less straightforward than that of insects and several different systems have been suggested. Krantz[174] remarks 'Acarology is, in fact, going through a state of systematic turmoil similar to that experienced in the field of entomology nearly a century ago.' One cannot, then, place complete confidence in an evolutionary tree based even on the recent classification which he recommends. Two other related difficulties are encountered in trying to convey a simple picture of the relationships of mites. Firstly, the major diagnostic features are difficult to discern and certainly not visible to the naked eye. For example, the group names prostigmata, mesostigmata, etc. refer to the position of spiracles on the mite's body. Secondly, these main groups are mostly much less consistent and easily recognized than the insect orders. This is not to say that many of the minor groups (superfamilies) do not have similarities in general appearance, especially to the expert. But only a few types are familiar, even to the entomologist, who is fairly used to looking at small arthropods; and, as one might expect, common names equivalent to moth, grasshopper, or flea are rather few.

Evolution of the Ectoparasites

I have sketched the outline of the evolutionary tree of terrestrial invertebrates and will now turn to the branches bearing the parasitic groups affecting man and trace the development of the relevant twigs and leaves. In Chapter 1, I pointed out that these parasitic forms have not been very successful, if we judge them by numbers and variety, as compared with free-living forms. This is perhaps to be expected, since their sources of food are more restricted and specialised. Furthermore, because of their parasitism, their bodies have become 'degenerate', in the sense that they tend to lose wings and become purblind. The long evolutionary history which resulted in a highly mobile flying adult, seems to have been in vain. Among the three parasite insects under consideration, there are varying degrees of close association with the host. Bed bugs live and breed away from the host's body, though all stages take the same blood meals. In the fleas, the adult is highly modified for living on or near the body, but these insects have inherited the distinct, free living larva; this develops in debris near the host's sleeping place. The closest adaptation has been possible in lice, which live their entire life cycle (including the egg stage) on the host's body. The itch mites, too, have become specialized skin parasites, though their 'degeneration' is less immediately obvious, because all mites have a humble, cryptic way of life which does not demand much in the way of locomotion or organs of higher sense (sight, hearing). Nevertheless, compared with free-living forms, the itch mites show distinct atrophy of their legs.

We shall probably never know how the original parasitic association with higher animals arose; but some clues can be gained from the nearest free-living relatives of the parasites, as we shall see later. Speculations as to when these adaptations occurred can only be based on the kinds of host affected and sometimes on their distribution in remote parts of the world. It will be remembered that a great deal of evolutionary development of arthropods had occurred before the mammals and birds appeared on the scene. Their emergence provided a tempting new environment for potential parasites to exploit. The earliest mammals arose in the early or middle Triassic (say, 200 million years ago) and the first birds somewhat later, probably in the first half of the Jurassic. Patterson[234] remarks: 'The largest known Mesozoic mammal . . . was no bigger than a cat, the great majority were comparable in size to shrews, mice and rats. So far as can be judged . . . all were associated in the small, terrestrial vertebrate community that evolved beneath the shadow of the dinosaurs. Under such circumstances, chances for the transfer of parasites among them would have been almost continuously available. Since a number were in all probability at least partially arboreal, opportunities for transfer from or to birds would not have been lacking.'

We must visualise then the widespread parasitism of these, as yet not very diversified mammals and birds by the original lice, fleas and perhaps skin mites. In the course of succeeding eons, the vertebrate hosts evolved into widely different types. To a considerable extent, this isolated their parasites, causing speciation and further evolution as would be expected from geographical isolation. Since the parasites became closely adapted to their hosts, the evolution of the two forms was co-ordinated; but the parasites would usually lag behind the hosts and their morphological similarities may sometimes provide a clue to the original unity of diverging hosts. Conversely, a great deal can be learnt about the evolution of the parasites from their host associations, which is very fortunate, since their fossil evidence is so exiguous. The subject has attracted much attention from parasitologists and it provided the theme of a Symposium in the University of Neuchatel, in 1957.[64,146,234]

Unfortunately, there are many difficulties in the study of co-ordinated evolution of parasites and hosts. In the first place, not only are there gaps in our knowledge but errors due to accidental infestation of domesticated animals or animals in zoos, or perhaps, contamination of skins in a museum. Secondly, as one might expect, the parasites are not completely specific. Lice and itch mites are, indeed, very faithful to a particular host, because their opportunities of transfer to new individuals are largely restricted to those of the same species; during copulation, nursing of the young or communal roosting of gregarious species. Even with these closely adapted parasites, there must be occasional transfer to new types of host; for example, to predators from their prey, in adjacent roosts or lairs, in dust-baths used by birds, etc. Such transfers are quite likely to happen in fleas, which spend only part of their life on the host's body.

There are further complications in the host-parasite relationship due to extinction of some species of parasite in the long course of evolution and infestation of the animal by a different species. Above all, there is the confusing effect of 'convergent evolution', which gives a specious similarity to animals which come to live in the same ecological niche, even if they are not related. Thus, parasites which live on the bodies of animals and birds all have the problem of moving easily through fur or feathers and at the same time, they must avoid being scratched out by

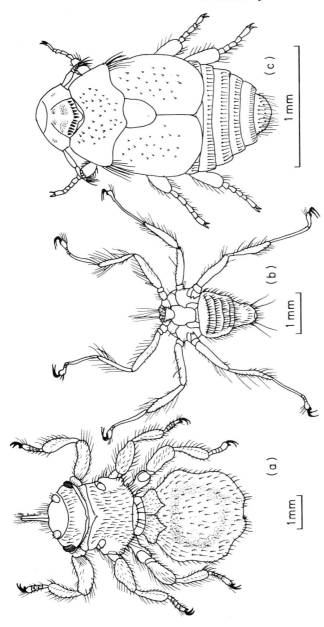

Fig. 2.2. Three unusual ectoparasites, to show bristles and combs. *a*, *Melophagus ovinus*, the sheep ked; *b*, *Penicillidia jenynsi* and *c*, *Platypsyllus castoris*. *a* and *b* are wingless parasitic flies. living on sheep and bats, respectively. *c* is an aberrant beetle, living in the pelt of beavers. (*a* and *b* after Imms;[263] *c* after Bugnion and Buysson, *Ann. Sci. nat. Zool.* (1924), **7,** 83.)

claws, beaks or teeth. Their response has been to develop strong, backward directed bristles and sometimes comb-like rows of cuticular teeth. These are particularly characteristic of fleas; but they are also found on Nycteribiidae (queer, wingless flies, parasitic on bats), Polyctenidae (bug-like parasites of bats) and *Platypsyllus castoris* (a peculiar blind beetle, living in the fur of beavers) (see Fig. 2.2). Among various mammal ectoparasites there is even a correlation between the spacing of their spines and the average diameter of the host's hair.

These brief digressions into the immensely complicated arguments about host-parasite relationships may form a background to the evolution of human ectoparasites, involving various warm-blooded animals and finally man. The types with which I am concerned vary considerably in their degree of isolation from free-living forms. Thus, at one extreme are the fleas, a type of insect without any close relations, while, in contrast, the bed bugs constitute a small family within a very large group, to which they show distinct affinities despite their adaptations to parasitism.

The Pre-historical Evolution of Bugs

The bed bug belongs to a huge and diverse order of insects known as the hemiptera, with about 60 000 species. The name, like those of most insect orders, refers to the wings; rather ineptly in this case, since the outstanding common character of the group consists in highly modified mouthparts, for piercing and for sucking sap from plants, or blood from animals. It is difficult to imagine the evolutionary changes by which tooth-like 'mandibles' and palp-like 'maxillae' were transformed into long stylets, with concave blades capable of forming a sucking tube when held together. Amazingly enough, the same transformation must have occurred quite separately in other groups of insects (e.g. fleas, mosquitoes); but in the hemiptera it must have been an extremely early development which has been retained while the group differentiated into an immense variety of forms.

Almost certainly, the original adaptation was for feeding on plants, which is the habit of the vast majority of the order. The earliest divergence was into two large groups, one of them with two pairs of membranous wings and one with the first pair of wings partly leathery and protective. The first group contains cicadas, leafhoppers, scale insects and a host of other plant pests; but it is the second group with which we are concerned. This separated into three main sections: the land bugs, the water bugs and the pond skaters (living on the water surface). Again we must narrow our attention to the land bugs, which comprise a large number of different families, including many plant bugs, which can be serious agricultural pests. Within this assembly, there are two groups of families which contain blood-sucking bugs. One family is the Reduviidae or cone-nose bugs of the tropics; the other is the Cimicidae or bed bug family. It is most likely that the blood-sucking habit developed quite independently in the two groups, probably from forms which fed on other insects, which they pierced and sucked dry. Many such predatory relatives exist today and some of them will prick the human skin if handled. The cone-nose bugs have not evolved far from their plant-feeding ancestor which they resemble in general appearance.

The minor branch containing the bed bug family (Cimicidae) contains two other related families; one of these is rather less uniform and specialized, the other perhaps more so (Fig. 2.3). The first is the Anthocoridae, a small family of about 300 species, mainly little flattened insects, which live in trees and among vegetation, preying on various small creatures. Some species have normal wings, but others have lost all but small rudiments of the forewings and some of these strikingly resemble the bed bug. One species, *Lyctocoris campestris,* is often found in food stores, barns and cottage roofs, living on insects there; this is one of those which may inflict a sharp prick with its mouthparts, if handled. The other relatives of the bed bug family belong to the family Polyctenidae and are all parasites, living permanently in the fur of tropical bats. Perhaps as an adaptation to this life, they have become viviparous, the nymphs being born at an advanced stage and requiring only two moults to reach adulthood. They are queer, blind, wingless insects; small, flattened and covered with spines and bristles.

The family Cimicidae is small (74 species) but well defined in the sense that despite some similarities of the related families just described, members of the bed bug family are really quite distinct and show considerable resemblance to each other. The anatomy, physiology, ethology and systematics of the group are very well described in an excellent monograph by Usinger and others.[308] They note that eight genera are restricted to the Old World and twelve to the New, while *Oeciacus* and *Cimex* occur in both. Most of the Old World forms are parasitic on bats, this being probably the original habit, while about half

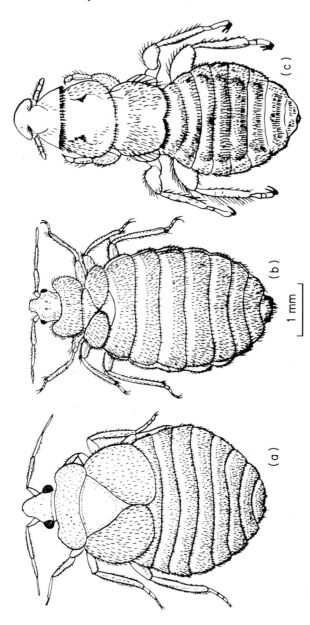

Fig. 2.3. The common bed bug and two distant relatives. *a*, *Astemmocoris cimicoides*, family anthocoridae (after a photograph by Carayon and Usinger). *b*, *Cimex lectularius* (after Usinger[308]). *c*, *Eoctenes nycteridis*, family polyctenidae (after Ferris and Usinger, *Microentomology*, 1939, **4**, 1)

of the New World genera, and two Old World forms, have become adapted to feeding on birds. The birds concerned mainly nest in cliffs or caves (e.g. swallows) so that their habitats resemble those of cave-haunting bats and one can imagine how the host transfer occurred. Most of the species of *Cimex* feed on bats; but the two species which have become adapted to man, show signs of extending their range to birds, as we shall see.

The origin of the association of bed bugs with man has been considered by many experts. It is generally agreed that bed bugs were originally parasites of bats and lived with these animals in tree holes and in caves. The cave habitat must often have been shared with man, especially during the last Ice Age in the northern hemisphere, about 35 000 years ago. It is in caves that bugs probably began feeding on man. The next question concerns the region where this happened; and we will consider first the common bed bug, *C. lectularius,* which is now cosmopolitan, having been carried all over the world by man and his possessions. In contrast to human lice and fleas, this widespread distribution is comparatively recent. In classical times, the bed bug was well known in the Mediterranean, and, as we shall see later, there were a number of references to it by Greek and Latin authors; but it did not spread to northern Europe until much later. Thus, Kemper[167] recorded the first mention of bug infestation in various countries, as follows: Germany (Strassbourg), eleventh century; France, thirteenth century; England, 1503; Sweden, later than 1807. However, there seem to be some errors here. The date for England is founded on a misprint (described later) and should read 1583. Secondly, the authority for Sweden (Fallen)[86] states that bugs had not yet appeared in the *interior* of the country (*In parte Scaniae meridionali nondum aparuit*). Fallen actually gives a Swedish common name, '*vaegglus*', for the bug, which suggests that it was generally well known. Peter Kalm[163] also remarks in his American journal (of December 1748): 'the people here could not bear the inconvenience of them any more than we can in Sweden'. He did not, however, find them prevalent among the Indians though the latter were plagued by fleas. It is believed that the most likely origin of the common bed bug was in the Middle East.[309] In this area were some of the oldest human cultures, providing an opportunity for the bug to leave the caves, with man, and infest dwellings in village settlements, which were later to grow into cities. About this time, the climate in the Middle East became drier so that the forest-living bats must have begun to vanish. But the bed bug was pre-adapted to take advantage of the new habitats provided by human dwellings. A few colonies must have remained associated with bats in caves and it is claimed that some may even yet persist.

The original association of tropical bed bugs (*C. hemipterus*) with man, is not so clear. It appears to have originated somewhere in South-East Asia, in which area *C. lectularius* is rare.[309] From this region, the tropical bug has been transported to tropical regions of Africa and the New World, though it is uncertain when this happened. It has not been able to invade temperate regions, though the more viable and prolific common bed bug has readily colonized many tropical countries. The two species do not interbreed and do not exist in mixed infestations. They have evolved in regions sufficiently far apart and for long enough to have reached a stage of reproductive incompatability. True, the two forms will copulate together; but the sperm fluid of *C. hemipterus* is toxic to *C. lectularius*. The opposite type of cross results in a few eggs, but the nymphs do not develop. Both species are less restricted to a particular host than other bugs and they have been found feeding naturally on bats (possibly a survival of the original habit) and on chickens, as well as on other animals associated with man, such as those in zoos. In the laboratory, they can be satisfactorily reared on the blood of various small mammals and birds. In one case, there seems to have been a natural evolutionary deviation to a new host. The pigeon bug, *C. columbarius* of Western Europe is closely similar to the common bed bug, to which it is very closely related. It is possible that this form became adapted to domesticated pigeons and doves in Roman times; alternatively, the original form parasitic on bats separately adapted to pigeons in this region. Whichever explanation is true, the isolation due to separate breeding sites has been maintained for long enough to result in partial genetic isolation.

The Pre-historical Evolution of Fleas

The fleas constitute a queer isolated order of insects, with some 200 genera and nearly 2000 species. The adults have evolved into a peculiar shape adapted to their parasitic habits, so that, as Snodgrass[290] puts it 'no part of the external anatomy of the adult flea could be mistaken for any other insect' (Fig. 2.4). Their mouthparts, too, are extraordinary in combining characters found in such different groups of insects as

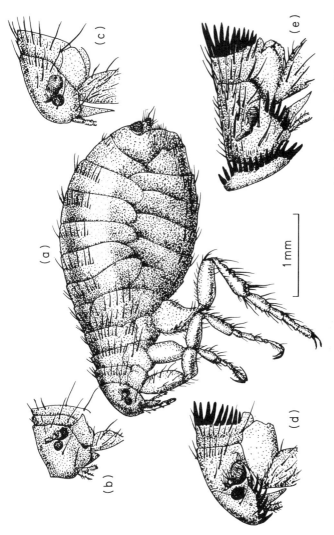

Fig. 2.4. *a*, human flea, *Pulex irritans*, surrounded by heads of some other fleas; *b, Tunga penetrans*, the jigger flea of the tropics; *c, Xenopsylla cheopis*, the tropical rat flea, vector of plague; *d, Ctenocephalides felis*, the cat flea; and *e, Stephanocircus dasyuri*, an Australian helmet flea. *P. irritans*, modified, after Hopkins and Rothschild;[147] others, modified, after Jordan, in Smart (*Insects of Medical Importance*, British Museum (Nat. Hist.), London (1948)).

booklice, thrips and bugs 'with an exaggeration of a structure otherwise peculiar to the Hymenoptera' (ants, bees, wasps, etc.). Fortunately, larval fleas and, to some extent, the internal organs of the adults, give some clues to their origin. On these grounds, it is generally agreed that they originated from a group of related forms described as the 'panorpoid complex' by the Australian entomologist, R. J. Tillyard, some 50 years ago.[133] The name is derived from *Panorpa,* the scorpion fly, which resembles the ancestors of the group more closely than the other living forms, judging from fossils in the lower Permian strata. From these progenitors, the other groups probably branched off as follows:

Moths and Butterflies (Lepidoptera, 160 000 spp.)

Caddis flies (Trichoptera, 3600 spp.)

Ancestral type ⟵ Modern Scorpion flies (Mecoptera, 300 spp.)

Fleas (Siphonaptera, 2000 spp.)

Midges, gnats, flies (Diptera, 90 000 spp.)

It is difficult to be sure when the fleas adopted the parasitic way of life and the path of the anatomical specialization this involved. We have noted that mammalian hosts have probably been available to parasites like fleas for some 200 million years. In 1970 a fossil was found in the lower Cretaceous of Australia which may possibly be a primitive flea;[266] if so, these parasites could have taken advantage of the new hosts very early. In any case, by the middle Eocene, 40 to 50 million years ago, fleas had evolved to their modern shape. This is proved by numerous specimens trapped in the Baltic amber,[240] which are excellently preserved and could not be mistaken for anything but fleas by anyone with rudimentary entomological knowledge. Further than this, the evolutionary isolation of the fleas, their wide dispersion, with small families isolated on ancient continents and adaptation to innumerable diverse habitats, indicates a long history of independent existence.[141]

A wide variety of different mammals have become regular hosts of fleas, though a few orders seem to have escaped, notably the odd-toed ungulates, the elephants, aardvarks and the primates. Relatively few fleas (about 100 species, in scattered genera) have become secondarily adapted to birds, especially various sea-birds, swallows and other passerine forms. The arbitrary nature of the flea's selection of hosts reflects their ecological convenience as well as their evolutionary history. One of the most important factors is the nature of the host's retiring place, because this is where the larvae of the flea develop, among the debris. For this reason, mammals that frequent nests, dens, holes and caves are especially attractive to fleas, though semi-aquatic forms, such as otters, beavers and water rats tend to escape. Conversely, wandering animals like ungulates, carnivores and primates are largely devoid of fleas. The exceptions are home-making forms: pigs among hoofed animals, foxes and badgers among carnivores and man alone among the primates.[146] Different fleas vary in their degree of dependence on a particular host. Many will feed occasionally on a variety of warm-blooded animals, but will only proliferate in association with one or perhaps a few special hosts. In some of the more frequently infested orders of mammals (insectivores, bats, rodents), there are almost as many species of fleas as there are of the hosts; and it would be expected that the diversification of the flea forms would have been parallel to the evolution of the mammals. Unfortunately, it is not very often possible to interpret the evolutionary relationships of the hosts from those of the fleas. In at least 16 different instances they have made the fundamental switch from mammals to birds, while fleas from marsupials will readily transfer to rodents. Furthermore, the anatomical peculiarities of the fleas are frequently modified by convergent evolution; that is the adoption of similar forms due to a common environment rather than inheritance from a common ancestor. There are numerous fascinating examples among fleas, which have been pointed out by various authors and extensively discussed in a recent monograph by Traub.[303] Among the characteristic types, which may be developed by unrelated families of fleas, are nest-fleas and fur-fleas. The former spend little time on the host, feeding mainly when it sleeps, so that they do not need extensive development of combs and bristles. Also, since the host is so accessible, they do not need eyes or great leaping powers, which indeed would be dangerous for fleas in the treetop nests or lairs. Thus they become weak, long-legged crawlers, purblind and lacking the powerful springing apparatus described earlier (p. 16). Fur fleas, however, spend much time on the host and are well supplied with combs and bristles. Even these show convergent adaptation. Unrelated fleas and some arboreal mammals have long pointed bristles, in contrast to the shorter blunt-toothed bristles of the fleas on terrestrial hosts. There are other adaptations of this kind, apparently related to environment but in an obscure way; thus, fleas of animals living in deserts tend to have combs of long, fine teeth closely set at the base. Again, combs of mole fleas are generally broader and blunter than those of shrews.

Many of the family to which the human fleas belong are devoid of combs; but they are by no means nest flea types and include vigorous leapers. Some are 'stick-tight' fleas which spring on to a host and then settle down for long periods in one spot, anchored either by powerful mouthparts or by a crown of spines in front of the head. This habit is most extremely developed in the jigger fleas of the tropics, the female of which burrows into the skin and develops vast numbers of eggs, so that she swells up to the size of a pea, with the tip of the abdomen projecting so that eggs can be ejected.

The so-called human flea has been found naturally infesting a considerable variety of animals, which appear to have little in common, except that they are all moderately large. These include canidae (fox, wolf), mustelidae (badger, skunk, pine marten), some large rodents and even deer and, especially, swine.[146] Many farmers in different parts of the world will confirm the liability of domestic pigs to support heavy infestations. Perhaps this is how the association with man began. As Holland[142] points out, the human dwelling, especially in its primitive form, is a reasonable substitute for a pig-sty! (We may also note that man and pig share several other parasites.)

In recent years, the standards of housekeeping in most civilized countries have improved to such a degree that *Pulex irritans* as a human parasite is becoming quite rare. A considerable proportion of complaints of flea bites (which I received over a 10-year spell as an advisory entomologist) could be traced to cat or dog fleas. The hygienic condition of the basket or kennel in which pets sleep may still be neglected to the extent of allowing a flea infestation to become established.

The Pre-historical Evolution of Lice

There are two kinds of lice: Mallophaga or biting lice and Anoplura or sucking lice; they are fairly closely related, so that some authorities put them in the same order, Pthiraptera. All of them live as permanent parasites on the bodies of mammals or birds. It is generally agreed that their nearest non-parasitic relatives are the Psocoptera or book lice (Fig. 2.5). These are small or minute insects with soft, rather broad

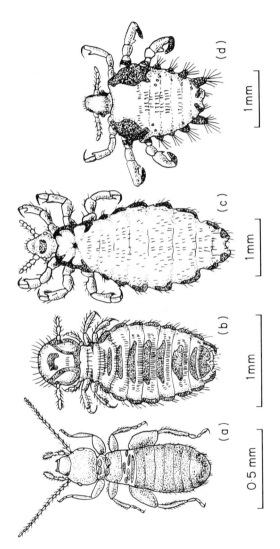

Fig. 2.5. *a*, A book louse, *Liposcelis granicola* (after Broadhead and Holby, *Discovery*, London (May, 1945)); *b*, the biting louse of the goat, *Trichodectes caprae* (after Evans, in Patton): *c*, the human body louse, *Pediculus humanus* and *d*, the crab louse *Pthirus pubis* (*c* and *d* after Ferris).[88]

bodies, some with delicate membraneous wings. The wings are lost in many species, however, probably as a result of their adopting a retiring habit, living among debris, under bark or among growths of algae or lichen. They live on fragments of vegetable or animal matter or on moulds. Several species live in bird nests and it is probable that from such a site (or perhaps a mammal's lair) the earliest ancestors of lice began their parasitic association with vertebrates. The most primitive lice were certainly biting forms (with chewing mandibles) which fed on fragments of skin or plumage. Almost certainly their original hosts were birds and their present wide distribution in that class suggests that they became parasitic on them at an early stage of their development and evolved with their hosts. As we have noted, however, evolutionary relationships are sometimes confused by the convergent evolution. Lice on the head and neck of birds tend to be rounded forms with large heads adapted to movement on the short feathers of this region. Being out of reach of the bill, they do not need to have the flattened form and rapid scurrying movements of forms found on the back and wings.[64]

In the course of time, the parasitic habit became more demanding, some forms taking sebum and perhaps serum by probing into quills or hair follicles. Later they began to take blood, either from small wounds or by biting into feather papillae. At some stage, mammals were invaded and a considerable variety are now parasitized by these biting lice. On the mammals, a further development took place; the biting mouthparts evolved into piercing stylets and the insects began to feed exclusively on blood. It is very difficult to say when this happened. It is noteworthy that seals are generally infested, and they have not had much contact with other mammals since their origin in the Lower Miocene (or late Eocene) 'except for momentary and fatal encounters with Polar bears'.[146] It is, however, conceivable that the sucking mouthparts were evolved separately from the Mallophagan types on the seals.

A tentative family tree of the lice may be constructed from the deductions of Hopkins[145] (especially the Mallophaga) and Ferris[88] (for the Anoplura). To some extent, the relationships are based on comparative morphology; but Hopkins, in particular, infers much from the host relationships, since lice are very conservative parasites. Often, they will starve to death rather than feed on blood of a strange host. Therefore, some of the evolutionary history of lice can be guessed from that of their hosts, which may be known from fossil evidence. On the other hand, there have evidently been some exceptional transfers to unrelated hosts. For example, there is a family of biting lice exclusively parasitic on Australian marsupials, with one exception, a kangaroo parasite which has transferred to dogs (presumably via the dingo). Again, sucking lice of the genus *Pediculus* are human parasites; but some are found on South American spider monkeys, which are not even close relatives of prehistoric man. It has been suggested that this transfer occurred in comparatively recent times from the habit of South American Indians of keeping such monkeys as pets.[145] A different transfer was accomplished experimentally about 20 years ago when, to avoid the unpleasant necessity of feeding human lice on themselves for experiments, some American scientists selected (with some difficulty) a strain which will feed on rabbits.[72]

It will by now be evident that lice are much more closely associated with their hosts than bed bugs or fleas and the species adapted to man are no exception. It is quite unlikely that the two genera concerned were casual transfers from animals sharing a shelter or dwelling with our ancestors, and virtually certain that they are parasites inherited from sub-human stock. As Hopkins[146] remarks: 'Distantly related species of *Pediculus* occur on man and the chimpanzee, and the genus is so widely separated from any found on non-primates that its occurrence on the Primates must be ancient—possibly of Miocene date'.

At some early time, too, human progenitors must have been invaded by the other genus, *Pthirus;* alternatively, the two genera, differing so greatly in proportions, may have evolved from a common stock on hominids, isolated merely by the region of the body infested. *Pthirus* appears to be restricted to primates (the only other species being found on gorillas).

The existence of two or more related forms of lice on the same host is not exceptional; thus, the genus *Linognathus* occurs on most (if not all) *Bovidae* and is commonly represented by two species, of which one is shortheaded and the other long-headed. The domestic sheep may be infested with long-headed *L. ovillus* (mainly on the head and body) and short-headed *L. pedalis* (exclusively on the legs). At the point when our human ancestors lost their body hair, except for the widely separated head and the pubic and axilliary patches, the sites of infestation were sufficiently distinct to discourage interbreeding. It also happens that the hair of the head differs considerably from that of the pubic region, which is much coarser and more widely spaced apart. Hence, the claws of the pubic louse *Pthirus pubis* are adapted to coarse hairs, and a sedentary life, which keeps them largely confined to that region. Oddly enough, there are occasional infestations of eyelashes which are similarly coarse and widely spaced.

When man began the habit of wearing clothes, a new ecological niche became available. One must suppose that the pubic louse had become too specialized and sedentary and that the colonisers were adventurous head lice. The new environment made available large areas of the body for feeding; but in other respects, it was less favourable. In the virtual absence of body hair, the lice were forced to crawl about on the garments, the fibres of which (wool, fur, cotton) would have been not too different from human hair. Experiments of a Canadian entomologist[138] have shown that the claws of the body louse cling tightly to fibres of this size, though they are less efficient at gripping the coarser fibres of certain synthetic textiles. Unlike human hair, however, garments are liable to shift about the body during daytime activity of the host; furthermore, they could be discarded temporarily. For these reasons, opportunities for feeding were more restricted than for the pampered residents of the hair (and beard) which could feed at any time. Probably for this reason, the body lice form has become distinctly more robust and resistant to starvation. Full separation into genetically isolated species has not yet developed, because double infestations have provided opportunities for interbreeding. Nevertheless, there are signs that isolation of the two forms tends to have increased with the progress of civilization, since the two forms have tended to become isolated to different sections of the human population. Head lice are characteristically found on young people, mainly girls, often quite clean and normal in most ways. Body lice, however, are restricted to people who rarely wash their underclothes and usually sleep in them (usually vagrants, in civilized societies) and infestations are common in older people. Whether both forms will persist in isolation for long enough to become separate species is open to question.

As a result of this incomplete separation of the head and body forms of *Pediculus,* there have been persistent arguments as to their systematic status. The type species, *P. humanus,* was described in the 10th edition of Linneus' *Systema Naturae* in 1758, but later in his *Fauna Suecica* of 1761, he referred to the head and body forms as distinct varieties. In 1778, De Geer named these varieties *P.h. capitis* and *P.h. corporis* respectively, and at various times subsequently there have been arguments about the status of these two forms. Are they, in fact, separate species, sub-species or mere varieties of a single species? This is a subject of which I made an experimental study some years ago,[49] so that I may venture some observations.

Body lice were obtained from naturally infested tramps, who assisted us with our researches for a small financial reward. There was no difficulty in maintaining colonies in captivity using the little gauze-bottomed pill boxes recommended by my former chief, the late Professor Patrick Buxton. These boxes, worn under one's socks during the day, enabled the lice inside to take a meal of human blood at any time. Head lice, however, taken from infested children, proved more difficult to maintain. Unless they were worn on the body continuously, a substantial proportion died during development. The problem was solved by wearing the small rearing boxes under crepe bandages round my ankles. Careful examinations and measurements of the two strains of lice, confirmed the observations of earlier workers in regard to certain differences in anatomy and size. For example, the body lice were, on the average, larger in all dimensions; but the differences overlapped, so that an occasional large head louse exceeded the size of a small body louse. Thus, while one could easily distinguish populations of the two forms, one could not be certain of every individual.

This work allowed me to re-interpret some earlier observations which had somewhat confused the issues. At the end of the First World War, Keilin and Nuttall[165] had claimed that head lice, worn on the body for a few generations, lost their head lice characters and 'turned into' typical body lice. My head lice, however, maintained their distinctness after nearly two years (45 generations) worn on my ankles. One must guess that the strain used by Keilin and Nuttall may have been a mixed race, in which case the stringent conditions of rearing in pillboxes would eliminate the less adaptable head louse forms by selective mortality.

Finally, there is another complication to be faced. Certain authors have suggested that diverse species or sub-species of lice occur on the various races of man. For example, *Pediculus nigritorum* on Negroes (Fabricus, 1805), *P. humanus chinensis* from Chinese (Fahrenholz, 1916) and *P.h. americanus*, from American Indians (Ewing, 1926). Ferris[88] has carefully reviewed all the evidence and dismissed most of this lavish proliferation of nomenclature. He bases his ideas on the current views of the history of man, which refers all existing races to the single *Homo sapiens*. It appears that the main racial types would probably have progressed through sub-specific to full specific identity, had they remained indefinitely in isolation. But the widespread migrations of man have resulted in interbreeding and gene exchange, which has reversed this process, though, in isolated areas, types exist which indicate the ways in which the sub-species would have differed. Ferris suggests that the isolation followed by re-mixing, would also have happened with the lice of man and so a similar situation is to be expected. Thus, a few typical head lice from natives in Africa or South America might differ as much as their hosts would differ in appearance from Europeans. But such forms are not genetically isolated and all shades of intermediates can be found, so that they can be regarded merely as races.

In conclusion, then, we may note that the future course of events is likely to blur geographical races of lice more and more; but the two forms on the body and head may tend to become rather more isolated.

The Pre-historic Evolution of the Itch Mite

I have previously stressed the minute size and ubiquitous presence of mites (reminiscent of micro-organisms) and the way in which various associations with other larger, animals have led to parasitism. In many cases, human beings can suffer from temporary or accidental infesta- tion with mites. For example, various grain infesting mites can cause dermatitis in people handling the infested products. Again, the house- dust mite, *Dermatophagoides pteronyssimus* causes asthma in sensitive subjects. From such more or less accidental associations, actual parasitism could evolve. The parasitic habit has developed sporadically in various groups, resulting in diverse types of parasitism, external or in- ternal. Professor A. Fain[85] has analysed various aspects of adaptation of parasitism in mites. Morphological adaptation can be regressive or constructive. The best example of regression is reduction in numbers of bristles and size of mouthparts, especially in the internal forms. The most obvious external modifications are the hooks, claspers and suckers developed in holding organs. Biological adaptations are noticed in the endoparasitic forms; notably a telescoping of the life cycle. Instead of laying eggs, they tend to become viviparous and stages are eliminated from the nymphal phase.

There are mites which have become external parasites of insects, millipedes, crabs, slugs and mussels. There are internal parasites of bees

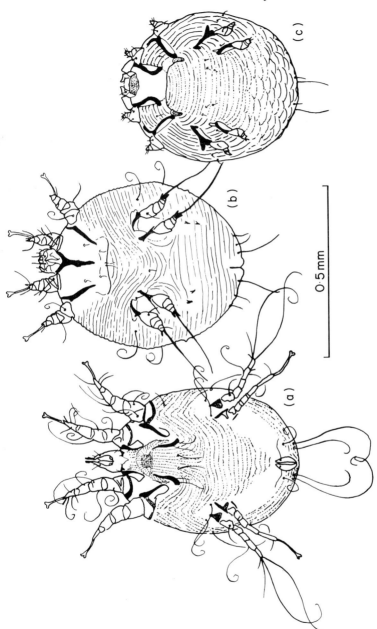

Fig. 2.6. Some parasitic mites. *a, Chiorioptes* Sp. (psoroptidae), a mange mite of domestic animals; *b, Sarcoptes scabiei*, the human itch mite; *c, Knemidocoptes mutans*, the scaley leg mite, affecting birds. (All after Krantz.)

0·5 mm

(such as *Acarapis woodi,* which invades the breathing tubes of honey bees, causing Isle of Wight disease), certain moths and even deep sea urchins. Parasites of vertebrates include such bizarre forms as the Cloacaridae, living in the mucosa of the genital openings of turtles and transmitted during mating of the hosts; also there are several forms which invade the lungs of respiratory tracts of various mammals, birds or reptiles. Ectoparasitic forms also occur in various groups; for example, the very tiny, worm-like demoditic mites, which cause mange in some animals. One species, *Demodex folliculorum* (Prostigmata) is common in the hair follicles of the eyebrows and forehead of man, but appears to be harmless. It is among the psoroptid branch of the Astigmata mites that the habit of ectoparasitism is most common. The superfamilies in the group include skin or fur parasites of mammals, feather or skin parasites of birds as well as groups invading sub-dermal or visceral tissues.

We have worked our way down to the close relatives of the itch mite and may note that several forms attack domesticated animals and cause mange (Fig. 2.6). This is the third group of mites responsible for this veterinary condition, which may be classified as demodetic, psoroptic or sarcoptic, according to the type of mite involved. Perhaps the best authority for the affinities of the sarcoptidae is Professor A. Fain,[84] who believes that *Sarcoptes scabiei* is essentially a parasite of man and closely related to a group of similar forms found on monkeys (genera *Prosarcoptes, Pithesarcoptes* and *Cosarcoptes*). Specimens closely related to *Sarcoptes scabiei* have been found infesting various domestic animals and also various other animals as different as carnivores, ruminants or marsupials. Fain, however, points out that he has never found examples in wild animals but only in captive ones or domesticated forms. The morphology of the species shows considerable variability in different populations and his theory is that all belong to the polymorphic species *S. scabiei* but some populations have transferred from man to animals in close contact with him and are in the process of becoming adapted to them.

3 The Harmful Effects of Parasites

In the last chapter I was speculating on the ways in which our four most intimate ectoparasites adopted their close association with man, in prehistoric times. Soon, I shall adopt a more anthropocentric role and consider various human observations and reaction to these creatures. But before leaving the more objective biological standpoint, it will be well to consider the mutual relations between parasites and man.

So far as the parasites are concerned, one might expect them to find the human body a kind of 'land of Cockaigne'; an almost inexhaustible supply of food and, for some of them, a warm equable environment. Most parasites however, run certain risks from their hosts, either from voluntary or involuntary reactions; and man is a specially dangerous animal for ectoparasites. These hazards are related, in various ways, to the harm which they do to man and they will become manifest as I discuss those.

From the human standpoint, the parasites are all more or less deleterious, though the reasons are somewhat different from many parasite–host relations of other animals. The direct drain on our vital resources, for example, is almost negligible. On the other hand, we have become much more conscious of the existence of ectoparasites with the increasing hygienic standards of civilization so that, uniquely, we have come to feel shame and disgust at their presence. A more tangible deleterious effect is the body's reaction to insect bites. Although this may be vexatious, on occasions, it is akin to the body's process of immunization and a kind of defence reaction against intruders. The most serious harm associated with insect parasites, however, is an incidental one, which can be quite as harmful to them as to us, that is when they become vectors of disease organisms. These are, then, four main reasons for objecting to ectoparasites: they feed on us, they are disgusting, their bites can be very irritating and they may carry disease. These subjects are worthy of a little closer attention; and since the last-mentioned was the most important, I will begin there.

Ectoparasites as Disease Vectors

Medical entomology is mainly concerned with arthropods which transmit disease. It is difficult for people in civilized temperate climates to realize this; indeed, the direct harm of biting and stinging insects may seem more obvious. Thus, in the United States of America,[232] more people die from wasp and bee stings than from poisonous snakes (rattlesnakes, cotton mouth, coral snakes, sidewinders, etc.); but the numbers are trivial (less than 40 a year). In the tropics on the other hand, the dangers from insect-borne disease are formidable. Despite great progress in the World Health Organization's campaign to eradicate malaria, there are still some 400 million people living in highly endemic areas, with nearly a million deaths annually. About 270 million people are afflicted with filariasis (disease due to parasitic worms, transmitted by insects) including onchocerciasis, which blinds one in ten of the sufferers. Other scourges include plague, typhus, relapsing fevers, sleeping sickness, encephalitides and Chagas' disease. Preventive medicine in the tropics demands the use of chemical pesticides to limit their spread, so that the health authorities there cannot easily subscribe to the banning of DDT and other life savers.

This kind of remark, however, is leading me into practical considerations of disease control in the modern world and away from the present speculations about disease transference by insects. To begin with, it must be realized that insects and other arthropods are invaded by various parasitic micro-organisms, just as we are; and equally, some of these organisms are malignant, others benign. The organisms concerned include viruses, rickettsia, bacteria, protozoa and helminths. In the past, the microbial parasites which are restricted to arthropods have been of little interest except to specialized scientists.[295] Recently, however, interest in them has greatly increased, since they offer a possible way of attacking pest insects by 'microbial control'. This is augmented by the current unpopularity of chemical pesticides and the development of insecticide resistant strains. A considerable proportion of these micro-organisms, however, involve man and his domestic animals directly, by alternating in parasitism between them and bloodsucking arthropods. The ways in which this came about provide a wide field for speculation. Presumably the remote ancestors of these microbes were free living forms, which then developed a parasitic association with one type of host (either vertebrate or arthropod), the alternating habit arising still later. This alternation between two very different kinds of host has allowed the parasites to take advantage of the useful features of each. The arthropods (which are mainly insects), provide an efficient means of wide dispersion. The vertebrate host, on the other hand, forms a large and long-living reservoir, capable of infecting many of the smaller, mobile 'vector' hosts. It is by no means easy to decide which of the two alternatives was the original host. We have seen that it is difficult enough to trace the evolutionary history of insect parasites because of the paucity of relevant fossils; and, so far as microorganisms are concerned, these are virtually non-existent. Sometimes clues can be found in the biology of related types; but often these are diverse and confusing. Thus, they may include free-living forms as well as parasites specific for vertebrates or for arthropods. It is somewhat easier in the case of the minute organisms known as rickettsiae (formerly thought to be intermediate between bacteria and viruses, but now considered to be much closer to the former). Virtually all of these at some stage are parasitic on arthropods, though some also invade vertebrates as alternative hosts and the inference is that the arthropods were the original hosts and the extension to vertebrates secondary.

A different type of evidence may arise from the degree of relationship of the hosts. Theoretically, the modern representatives of the original host should be more closely related that those of the secondary one. Thus, malarial parasites are all spread by mosquitoes, but involve quite diverse animals and on these grounds, Sir Richard Christophers[62] suggested that they were the original hosts and that the vertebrate hosts were an arbitrary tree-living assembly: birds, bats, monkeys and arboreal man. However, it is currently believed that the remote ancestors of the malaria pathogen were intestinal parasites of a vertebrate, probably an aquatic one. At a later stage, they invaded tissues beyond the gut wall and eventually the blood stream. This paved the way for transmission by blood-sucking ectoparasites; and the final specialization in one group (Plasmodiidae) involved only mosquitoes.

A quite different situation exists with the trypanosomes, which, according to the balance of modern opinion, began as invertebrate parasites.[137] The trypanosomes include species responsible for sleeping sickness, carried by tsetse flies in Africa, and Chagas' disease, transmitted by triatomid bugs in South America. Other trypanosomes, not involving man, are carried by leeches (to frogs or fish). Despite the variety of invertebrate hosts, there seems to be only one trypanosome passing directly between vertebrates (a venereal infection of equines, almost certainly a secondary adaptation).

Whatever the origin of two-host micro-organisms, it appears that considerable evolutionary adaptation was necessary to facilitate the double life cycle. Invasion of an alternative host must involve adaptations similar to those in the initial parasitism. These include the development of enzyme systems capable of digesting food sources in the new host's tissues, and also of combatting the protective immunological reactions of this host. These are complex biochemical matters which will not concern us here; but something may be said of the simpler mechanical problem of transferring from one host to another. There are various ways of doing this, of varying complexity. A very simple method is by mere contamination; for example, insects like houseflies and cockroaches are liable to visit human faeces (if they are accessible) for feeding and egg-laying and, after crawling over and feeding on such potentially infective material, they may visit human food and deposit germs on it. This kind of insect involvement necessitates a large element of chance and no disease germ relies entirely on it; fly-borne infections, for example, are commonly spread by contagion or in polluted water. One defect of the accidental contamination of insects is the short survival of pathogens. There would clearly be an advantage in a more permanent stay in the vector insect, by using it as a second host. This has developed in various ways, reaching increasing levels of complexity and efficiency. An important step was taken by the parasites invading the blood stream of the vertebrates, thus becoming accessible to blood-feeding insects. A simple transfer then could be made by contaminated mouthparts; but this again is inefficient, because of the very tiny quantity of blood involved. The next step was a proliferation of pathogens in the gut of the insect, which would eventually pass out infected faeces. These could re-enter a human or animal host by being scratched into a small wound, or by being inhaled. Alternatively, by proliferating in the front of the insect's gut, they could be reintroduced by the bite, the process being helped in some insects by partial blockage of the gut, so that they are forced to regurgitate into their next victim. A more intense parasitism of the insect vector would involve penetration from its gut into its body cavity; and eventually a cycle developed in which the pathogen made its way to the proboscis or to the salivary glands, ready to invade another host at the next feed. This final arrangement is responsible for transmission of a number of serious diseases, including malaria, yellow fever, sleeping sickness, filariasis, dengue and various encephalitides. It is highly efficient in allowing the insect to infect a number of new cases, as long as the parasite does not harm it.

On grounds of survival potential, a parasite is likely to evolve towards a benign state, since harmful effects, leading perhaps to the death of the host, would be an obvious disadvantage. Therefore, one would expect a degree of virulence to be an indication of more recent parasitism; but there are complications. The likely course of an extension of an animal parasite to man may be as follows. A well-established vector-borne disease of animals may, from time to time, become accidentally implanted in man, due to the chance feeding of an infective vector on him. At this stage, the pathogen may not be able to overcome the natural human immunity reactions, so that it never develops a heavy infection; consequently, its virulence would be low. Suppose, however, that a change in human habits or the environment allows the vector frequent opportunities to feed on man, so that these chance infections become rather frequent. A mutation is likely to arise which will allow the parasite to establish itself in the new host. Human defences are no longer able to suppress the parasite rapidly and the result is severe virulence at this stage. Finally, however, long established diseases (whether vector-borne or not) presumably evolve to a stage of lower virulence, with improved chances of survival of both man and pathogen.

While the involvement of man in infections transmitted by ectoparasites can be considered a recent occurrence in geological terms, it was clearly a long time ago historically. Probably many such adaptations occurred when man abandoned a nomadic hunting existence and settled in communities with domesticated animals.[65] He was thus in close contact with dogs and cats, horses, cattle and goats, not to mention rats and mice, sparrows and pigeons. As we have seen, this is probably the time when the bed bug and possibly also the human flea became closely associated with man. It is curious, therefore, that no specific disease is regularly transmitted to man by either of these parasites. On the other hand, important examples of disease transmission occur with human lice and also with certain rat fleas; and these deserve closer attention.

Typhus

The appalling disease known as epidemic or spotted typhus is caused by an organism called *Rickettsia prowazeki*. The name honours two early workers who died of typhus while studying the disease; the American, H. T. Ricketts and the Austrian, von Prowazek. Typhus is a serious disease, quite often fatal in the days before modern treatments, especially in older people, and it is always fatal to the human louse which spreads the infection. Accordingly, one may suppose that this is an example of the second stage of the extension of infection to new hosts, since a more completely evolved parasitism might be expected to be benign. The question arises then: whence did this dangerous parasitism arise? It is possible that the answer may be found in an organism related to that causing epidemic typhus; this is *Rickettsia typhi* (formerly *mooseri*), which has achieved a stable parasite-host relationship with rodents and their ectoparasites (fleas, lice and biting midges). Among the rodents which may become infected are domestic rats and, perhaps, mice. This may involve man, since certain fleas of domestic rats will bite man when they are very hungry; when the rat dies or is killed, for example. Thus, the rickettsiae may be transmitted to man, in a similar way to plague bacilli, which we will consider shortly. Unlike plague, however, it is not the insect's bite which is infective but its faeces, which may get on to the skin and enter an abrasion, when the spot is scratched. The result is a mild, febrile disease known as murine typhus. Should the sufferer be lousy, however, his lice will become infected from rickettsiae in the blood stream. These multiply in the louse's gut, causing considerable damage, which is finally fatal to the louse; but for a short time the louse is capable of infecting other human cases by its infective faeces. Sometimes this may happen in a community with low standards of hygiene and general lousiness, in which case an epidemic of murine typhus could well occur.

The dangerous spotted typhus, however, does not involve vertebrates other than man and the alternative host is, solely, the human louse. Experimentally, it can be shown that both head and body lice (and even the generically distinct crab louse) can transmit the infection; but epidemics have always occurred in conditions where body lice were particularly prevalent and this is the usual vector. Probably *R. prowazeki* has evolved from *R. typhi*, which it resembles in several aspects and at one time the two pathogens were believed to be mere varieties due to different environmental factors. It appears that, in quiescent periods, *R. prowazeki* remains as a sub-clinical infection in man, which may be exacerbated into a virulent form by conditions of stress. This appears to be the explanation for remissions of dormant typhus noted in European refugees to America at the turn of the century and then described as Brill's disease. Conditions of stress and famine together with overcrowding and general lousiness have been commonly associated with war, especially in the past few centuries. As a result, epidemics of typhus have occurred with immense loss of life. It is estimated, for example, that some 30 million cases of typhus occurred in Eastern Europe between 1917 and 1923, with some 3 million deaths. Zinsser[331] gives many examples, often with fascinating political consequences, from the history of the last millenium. It is interesting to note, however, that despite a review of pestilences described in classical times, he could find no earlier reliable description of typhus before an isolated and somewhat dubious instance in 1083 recorded in a monastery near Salerno. After a lapse of 400 years, an undoubted epidemic in Spain occurred, apparently brought there by soldiers returning from Cyprus. Following this there are repeated instances, reaching appalling frequency in the sixteenth and seventeenth centuries, especially during the Thirty Years' War. While large epidemics were associated with widespread famine and war, smouldering endemic typhus was common in filthy overcrowded prisons and the almost equally squalid ships of 200 years ago. Hence the disease was often known as jail fever or ship fever.[190]

From the rather scanty evidence available, Zinsser suggests that the mutation of rickettsiae to the form causing epidemic typhus may have occurred during actual historical times, say, about 1000 years ago. Descriptions of disease were notably unreliable in ancient times and epidemics were often composite mixtures of several infections (plague, dysentery, smallpox and so forth) which complicates the difficulty of identifying the earliest typhus epidemic. On the other hand, the poor adaptation of the rickettsiae to its new hosts—both lice and man—supports the theory of recent development. Cockburn,[65] discussing the evolution of disease types points out that epidemic diseases need population groups of a certain minimum size, whereas endemic infections can smoulder in small isolated populations. There were, of course, large armies with presumably the right conditions for epidemic typhus in classical times; but one must remember that events in classical times seem to us telescoped together from the large expanse of time over which they occurred. Probably large organized military expeditions, causing havoc in urban populations, were much more frequent during the later period when typhus became really well established.

Perhaps as an anticlimax it may be worth referring to a more successful parasitic rickettsia of lice and man, which does little harm to either. This is *R. quintana* which was the cause of trench fever in the first World War, a mild disease for men and quite harmless to lice.

Louse-borne Relapsing Fever

In parts of Asia, Eastern Europe and belts of Africa north and south of the Sahara, there are occasional epidemics of an undulant fever transmitted by lice. In the warmer parts of this range, the disease may be confused with malaria, due to the intermittent periods of fever. In cooler countries, it has been overlooked, because epidemics have sometimes coincided with typhus from which it was only distinguished with certainty about 1840. The organism responsible is a spirochaete, *Borrelia recurrentis,* one of a group of blood parasites causing forms of relapsing fever. Related species, transmitted by ticks cause similar diseases in various hot countries. The pathogens all look rather similar and, while immunological differences exist, they are not very sharp. The tick vectors are quite unaffected by them and can transmit them to their offspring via the eggs. Wild rodents, too, are not appreciably affected, so that a benign parasite cycle seems to have been established. The lice responsible for human infections are also unharmed, though they do not transmit the parasites via the egg. On these grounds, some authorities believe that the original parasitism was in ticks, later extending to vertebrates and finally lice and then man. On the other hand, others consider that the first double cycle involved small mammals and their ectoparasites and that ticks were involved later; from them, occasional infections of man occurred, as they do now. The next step would be the involvement of human lice, resulting eventually in a new species of *Borrelia* adapted to them. If, indeed, the original *Borrelia* was an invertebrate parasite, it was an anomalous member of the group, since other kinds of spirochaete are associated with ulcers of the mouth or genitalia, lung abscesses or tropical sores. Furthermore, distant relatives of these spirochaetes are the treponomas, responsible for syphilis, and the tropical skin diseases yaws and pinta. Still others are free living, in mud, sewage and contaminated water, while some are found in oysters and other molluscs.

The organism transmitted by lice, *B. recurrentis,* occurs in the blood of febrile patients and is sucked up by lice feeding on them. After about six days, the spirochaetes appear in the blood of infected lice in considerable numbers. They remain in the body cavity of the louse for the rest of its life; but they cannot escape and are not transmitted to other lice or to man, unless the insect is injured. If the louse's skin is broken, however, its infectious blood can escape. This must often happen when infested people destroy lice by 'popping' them between fingernails (or even by crushing them between the teeth, as has been observed in some primitive races!). Infection through mucous membranes or abrasions produced by scratching, produces the next human victim.

Fleas and Plague

Everyone has heard of plague. Although Western Europe has been comparatively free for several centuries and although it is now vastly reduced throughout the world, the intense horror of a plague visitation keeps its memory alive. Diverse books have been written on plague in general[136,249] or on particular epidemics, such as the Black Death.[331] Accordingly, the association of plague with rats and their fleas is reasonably familiar to educated people.

There are, indeed, still some mysterious circumstances about the disease; for example, it is difficult to speculate on whether the bacteria responsible (*Yersinia pestis*) are descended from parasites of mammals or of fleas. Not very much can be deduced from the characteristics of related bacteria. Only one of these, *Francisella tularensis*, is involved in a disease transmitted by arthropods (tularaemia) and this species differs substantially in morphology and immunology.

Consideration of the evolution of plague from the original hosts to the urban epidemics can be more rewarding. So far as we can judge, the original alternating cycle was between certain wild rodents and their fleas, a cycle which continues today. Yet the parasitism does not seem to have reached a stable benign condition, since many of the rodents suffer disease and there are periodic epizootics with many deaths. Furthermore, the mechanism of transmission from the flea to the mammal, seems curiously fortuitous, which does not suggest a mechanism perfected by a long history of evolution. Sometimes, it seems, when there are large numbers of voracious fleas attacking a plague-infested rodent, a considerable proportion may shortly attack another animal, especially if interrupted in mid-feed. In this way, plague bacilli may be mechanically transferred on their contaminated mouthparts; but this mode of transmission is not very efficient, because of the very tiny traces of infected blood carried by each flea, so that effective transmission is believed only to be possible by mass attacks. A more efficient mechanism of transmission is one unique to fleas. What happens is that plague bacilli proliferate in the flea's intestine and tend to form a gelatinous clump. In some fleas, this mass of bacteria forms a plug, which eventually blocks the gut, usually in a kind of gizzard lined with spines, called the proventriculus. The flea is then unable to swallow its next meal of blood; but as it gets hungrier, it makes repeated efforts to do so. The blood is sucked desperately up to the blocking plug but usually squirts back into the host animal. This often results in detachment of clumps of bacteria and a heavy infection of the animal concerned. For some reason, this blocking of the gut occurs sooner and more frequently in some species of flea than others; and these are the main flea vectors of plague. When blocking of the gut occurs, the flea invariably dies of starvation, commonly within a week. Some authorities believe that the bacillus has an additional toxic effect, though it does not invade the flea's tissues. On the other hand, those fleas which do not become blocked seem to live normally and may get rid of the bacteria. Since, however, the most efficient transmission mechanism by blocked gut is lethal to the flea, we have further evidence of the ill-adapted parasitism of the plague bacillus, in that it is so dangerous to both its hosts.

So far, we have recognized plague as a disease of rodents, especially ground living forms. The original home of the disease is thought to be central Asia, though another very ancient site is located in Central Africa. In this form, the disease is of little importance to man, though occasionally some hunters catch it when handling infected animals. From the wild rodents, plague may transfer to domestic rats in urban fringes where the two mix (or where there exist semi-domestic rodents to act as intermediaries). An epizootic among domestic rats is the normal prelude to a human plague epidemic. As the rats die, starving fleas leave their bodies and some of them (especially the tropical rat flea *Xenopsylla cheopis)* will feed on man.

In man, the infection is commonly confined to the lymphatic system at first and tends to be concentrated in swellings (buboes) usually in the groin or armpits. In this condition, a man is virtually not infectious and represents the end point of the disease. In fatal cases, a short period of general septicaemia may occur, but these do not involve prolonged hazard; and in any case, the human flea, *Pulex irritans,* very rarely gets a blocked gut, and so is an inefficient vector. The disease can, however, spread directly from man to man in cases where serious lung involve- ment develops; then there may be droplet infection of the highly lethal 'pneumonic plague'. Curiously enough, however, such epidemics are generally less extensive and self-sustaining than the flea-borne type. It was the flea-borne 'bubonic' plague (easily recognizable because of the characteristic swellings) which was prevalent in most of the historical epidemics.

The rodent chiefly responsible was the black rat (*Rattus rattus)* which combines foraging activities with intimate association with man, especially in primitive wooden buildings. The brown rat *(R. norvegicus)* is an earth-living form and while it takes advantage of human waste, it does not adapt well to living in human dwellings. At present, it is a com- mon pest of sewers. Mice, indeed, live close enough to man, but they are less susceptible to plague than rats; and furthermore, their particular fleas are unwilling to bite man. The black rat, on the other hand, is frequently infested with *Xenopsylla* fleas, which readily feed on man and which are very prone to develop blocked guts, if infected with plague bacilli.

We have now narrowed down to the prime causes of plague: black rats and tropical rat fleas. Wherever these two exist together there is the possibility of a plague epizootic among rats followed by a plague epidemic in the human population. The earliest known plague in historical times was, perhaps, the pestilence that struck the Philistines in 1320 B.C., which, as described in the Bible (1 Samuel 5 and 6) affected them with 'emorods in their secret parts'. Some controversy exists as to whether this refers to plague buboes or haemorrhoids; though an 'epidemic' of piles seems both unlikely and not worthy of such serious notice. A further epidemic seems to have occurred about 200 B.C. in Libya, Egypt and Syria; but the earliest well-authenticated and really extensive pandemic occurred in the fifteenth year of the Emperor Justinian's reign (A.D. 542). Over a period of some fifty or sixty years, many millions of people were affected, with great loss of life. The next intense visitation of plague, which involved both the bubonic and pneumonic forms, raged through Europe and a large part of Asia in the fourteenth century. This pandemic is known to every schoolboy as the Black Death, for reasons not entirely clear. (The victims do not turn black, as commonly supposed, though dark marks on the skin may be seen in septicaemic cases. Actually, there is no evidence that the term was used in England until the seventeenth century. Probably the name stemmed from a literal translation of *pestis atra* or *atra mors,* though here *atra* would have implied 'dreadful' or 'terrible' rather than 'black').[330] By the twelfth century, black rats had been introduced into Europe and must have been involved; though whether they were accompanied by the tropical plague flea at this time is difficult to say. At the present time, both black rat and plague flea are restricted to sea ports. In any case, it is estimated that some 25 millions, or a quarter of the population of Europe succumbed. Incidentally, it is rather interesting to note one source of such estimates is from ecclesiastical records, since replacement of country priests due to plague death would be recorded even when no such records of the general population were kept. In the centuries following the Black Death, the residue of the infection moved sporadically around Europe; but from the seventeenth century onwards, a decline set in, commonly attributed to elimination of black rats from improved house construction. Curiously enough, however, a simultaneous decline occurred in much of Asia, which is more difficult to explain. Towards the end of the nineteenth century, plague made a final deadly resurgence. From an epidemic in Yunan, a remote province in central China, it reached the coast about 1870. It fluctuated and finally broke out seriously in 1894 in Canton and Hong Kong, and thence spread round the world.

The original sources of these plagues were, presumably, the persisting foci of wild rodent plague in Central Asia, or perhaps in Africa. The spread from such foci in ancient times must have been slow; first to urban centres nearby and thence perhaps with rats or fleas accompanying caravans. In the nineteenth century, however, a sudden increase in the speed and extent of human transportation spread the 'seeds' of plague to the coasts of distant countries throughout the world, e.g. India, South Africa, Australia, South America, California. Local outbreaks occurred (usually in sea ports) and were duly quenched; but, in the meantime, the plague bacillus had spread into the interior and eventually had reached the wild rodents and their fleas in the rural districts. Here, new permanent fulminating foci were established and continue to this day. They represent a minor threat to hunters and a permanent menace to local towns, except where modern hygiene and improved housing have permanently eliminated the black rat.

Allergic Skin Reactions

It is common knowledge that insect bites are liable to cause irritating skin reactions. Since the bites of many insects are almost painless at the time and may occur during sleep, most people would be unaware of them were it not for the subsequent irritation. It is vaguely realized that this reaction is much worse in some people than others. Nevertheless, if a sensitive and non-sensitive person are equally exposed (sleeping in the same mosquito-haunted bedroom, for example) the former will ruefully claim that all the pests had chosen him as their victim and neglected his companion, which may not have been the case. Further than this, there is some recognition of the fact that sensitive people can become desensitized. At least, it is well known to people in tropical and subtropical countries that the newcomers suffer more from insect bites than old residents.

It is only in comparatively recent years that the nature of insect bite reactions has received much attention. Medical entomologists had been largely preoccupied with the role of arthropods as disease vectors, while dermatologists had been interested in more serious skin diseases. It is now generally recognized[205] that the skin reactions after bites are a form of allergy; that is to say, a kind of aberration of the body's normal immune response to invading germs and toxins. Following an exposure to a foreign protein entering the tissues, the body after a little delay, manufactures an 'antibody' to combine with the invading antigen in order to neutralize it. Antibodies are made from a class of proteins called globulins and they are tailored to match the particular antigen present; hence, the reaction is rather specific. Neutralization may involve agglutination, lysis or digestion of the antigen-antibody complex by phagocyte cells.

Disturbances in this process can have more or less harmful effects on the organism concerned. Injection of a foreign protein followed by another when the antibodies have been prepared, can cause a severe general reaction known as anaphylactic shock. More localized effects occur when the antigen-antibody complex becomes attached to specific tissues: e.g. asthma when bronchial regions are affected or hay fever affecting the nasopharangeal region. Finally, a local effect, known as an

Arthus reaction, is caused by accumulation in the skin; and this is what can occur with insect bites (and stings). The local effects are believed to be due to release of physiologically active chemicals such as histamine, which causes reddening, swelling and irritation. Some relief may be obtained by using anti-histamine drugs. Attempts to cure the trouble can be made by giving tiny, but increasing doses of allergen in order to institute a normal reaction, i.e. de-sensitization. This may sometimes occur spontaneously as it often does, in fact, in regard to insect bites and stings.

In the case of insect bites, the allergen consists of minute traces of saliva injected during feeding. While it cannot be expected that all bite allergens are similar, some interesting American work with cat fleas has revealed some of the allergenic constituents of their saliva.[23] By inducing thousands of the fleas to feed on distilled water, through a membrane, their oral secretions were collected and concentrated. (One million 'flea-hours' yielded about half a gram!) Guinea pigs, sensitized normally to cat flea bites, were used to assess the allergenic property of the preparations.

The allergic reactions to insect bites does not, as might be expected, develop at the first attack but needs delay and exposure to further bites. The situation is complicated by the fact that there are two, quite distinct, skin reactions as follows: (i) An 'immediate reaction' consisting of a small flat white weal, surrounded by a red flare, which develops in a few minutes and subsides within an hour. (ii) A 'delayed reaction' with much swelling and reddening of the region, which may persist for days. Both forms are irritating, but the delayed reaction is far worse and more prolonged than the immediate one. The appearance and relative severity of these reactions change with the amount of exposure to insect bites and the degree of immunity which may eventually be reached. The usual course of events, from the first exposure to regular bites from a particular insect, are as follows:[205,215]

	Immediate Reaction	Delayed Reaction
Phase i	—	—
Phase ii	—	+
Phase iii	+	+
Phase iv	+	—
Phase v	—	—

It will be noted that the first response is the (severe) delayed reaction; then both forms; and next only the immediate weal. Finally, in some fortunate subjects, a state of complete immunity (phase v) is reached. At this time, the only sign of insect bites are tiny red marks at the sites of the bites.

Not all people reach a full degree of immunity; and since the reactions are rather specific, it is possible to develop to different stages for various biting insects. I, myself, am largely immune to most of these pests (though only to phase iv for some) which is fortunate, in view of my occupation. It seems that it is far easier to acquire immunity to some species than to others. Thus, over a period of about ten years while I was working on human lice, the creatures were fed from little pill boxes strapped round my ankles. While the reactions were not as severe as some I have seen on colleagues, I never became immune to the irritation they caused (an experimental study of reactions to lice bites was described by Peck *et al.*,[237] in 1943). On the other hand, I suffer no reaction from bites of bed bugs or fleas, though we normally feed both these insects on rabbits and seldom on ourselves.

The intense irritation experienced by sensitive subjects is unpleasant enough; but the inevitable scratching which follows may lead to secondary infection. Thus, impetigo in schoolchildren is a common result of persistent infestation by head lice. On the other hand, the irritating effects of insect bites are serious for the ectoparasites too, since they are often killed both accidentally in scratching and intentionally in delousing operations. This occurs in animals as well as man, it being often observed that heavily infested individuals are sick or undernourished, and the parasites flourish because weakness or apathy has decreased the normal self-grooming of animals and the preening and dust-bathing of birds. Ectoparasites generally have virtually unlimited food supply and few enemies of their own size, so it is their host who is the main check on their proliferation. Buxton[54] points out that primitive disinfestation measures are the probable reason for the small numbers of lice found on most lousy people, even under unsanitary conditions. The rare, excessively heavy infestations must occur in persons who have become completely immune.

Sensitization also appears to occur in cases of scabies, though here the antigen is not a bite but presumably some secretion or excretion of the mites in the skin. The reaction takes perhaps a month to develop and early experimental infections with scabies were found to be symptomless. When the sensitization develops, the patient suffers severe itching, which becomes almost intolerable at night. A characteristic rash is found on certain parts of the body, and the scratching induced by the irritation usually leads to secondary infections which can be quite severe. One beneficial effect is that the scratching probably destroys a considerable proportion of the mites and keeps their numbers low. The allergic condition remains, even if the infestation is cured and there is an immediate reaction to a new infestation of mites, which accordingly have some difficulty in becoming established. A cured case of scabies, therefore, possesses some immunity. It is conceivable that if the infestation is not cured, the host's sensitivity may decline (in analogy with immunity to insect bites). Mellanby[213] suggested that this might result in the very rare heavy infestations with mites known as Norwegian scabies, since none would be destroyed by scratching.

In conclusion, therefore, we can recognize that skin reactions to ectoparasites are a kind of defence against them. The insect bites irritate us into trying to get rid of them. The scabies reaction leads to mite-destroying scratches and skin conditions unfavourable to mite survival.

Loss of Blood, etc., to Parasites

So far as their actual food requirements are concerned, the ectoparasites of man approach the ideal condition of thriving without significantly affecting their host's tissues. This seems evident from the following calculations. An egg-laying female bed bug has the greatest food demand and takes about 8 mg blood at a full meal[160] and the average for all stages cannot be more than, say, 4 mg. They feed about once a week, under British summer conditions, so that a heavy infestation of about 1000 bugs will require 4 ml per week or about 0.6 ml per day. This is close to the daily requirements of a very heavy louse population which I estimate as about 1 ml. Thus, an egg-laying female takes about 1 mg, twice daily;[54] so, with a population mean of about half this, 1000 lice would take $0.5 \times 2 \times 1000 = 1000$ mg. Fleas have even smaller requirements, taking only about 0.5 mg blood at a meal, perhaps every 4 or 5 days.[17]

The effects of these losses on a full grown man are quite negligible, since he has about 5 litres of blood and can spare about a pint for transfusion without ill effects. A regular loss of over 10 ml per day might cause anaemia; and of course, proportionately smaller losses would be harmful to young children. Usinger,[308] indeed, states that after regularly feeding a bug colony for seven years, he developed a measurable degree of anaemia, which disappeared when he ceased. Nevertheless, it seems unlikely that a blood loss of 1 ml or less per day could be serious; and these figures correspond to very heavy infestations.

It is difficult to make similar calculations about depredations of the scabies mite. But since this feeds only on the horny layers of the skin and since the numbers are generally not excessive, it is most unlikely that the direct tissue damage is of any significance.

Normal Disgust and Neurotic Abhorrence

The later chapters of this book will be devoted to the history of human reactions to ectoparasites. But before embarking on this, I would like to consider the final state of feelings of civilized man to these parasites and the consequences of these feelings for both. Earlier sections of this chapter have dealt with various logical reasons why we should dislike these insects and mites; but our feelings undoubtedly go deeper than mere reasonable objections. There are, indeed, other household insect pests which do very little real harm but are regarded as highly unpleasant. Most civilized people regard even quite harmless little crawling creatures (such as earwigs or woodlice) with disgust, especially in their home; and they would certainly react strongly to bed bugs in the bedrooms, or lice on their persons. Most probably this is a cultural tradition rather than an instinct, for very small children or savage peoples are generally much less fastidious. Presumably we are conditioned by exclamations of 'Ehrr! Dirty!' from our mothers, when we innocently try to pick up crawling creatures in our babyhood. Such feelings of repulsion are quite healthy and akin to our abhorrence of refuse or excrement. It is largely upon such feelings that high standards of hygiene are based. Before DDT was introduced, the health authorities in this country were much concerned with bed bugs. My old professor, the late J. W. Munro, called this pest the 'Slumbug', pointing out that it was both a characteristic of bad housing conditions and an indirect cause, since only slovenly people would consent to live in infested houses, while the better tenants moved away. In addition to their generally disgusting appearance, bed bugs have two other unpleasant characteristics: they make a mess and they stink. The mess is due to their frequent excretion, which causes brownish, yellowish or black spots on the wall near the crevices where they hide. The dark marks are due to the presence of partly digested blood in their faeces excreted to make way for a fresh meal. Pale excrement however is more usual during periods of starvation when, as Wigglesworth[320] says 'the urine is in the form of a pultacious mass, which dries to a yellow powder'. Whatever the hue, the faecal marks form unpleasant speckling on wallpaper and furniture, and are good signs of the presence of the hiding bugs.

The stink of bed bugs is generally thought to be due to secretions of special glands, under the thorax in the adult forms. Those glands emit a curious odour when the bugs are disturbed; its purpose appears to be to alarm and disperse other bugs.[185] I once extracted some and offered it to an organic chemist to smell. He said that it was quite familiar, but he could't put a name to it! I do not think my chemist friend was living in bug infested quarters, because the gland odour has been shown to be due to rather ordinary chemicals: hex-2-en-1-al and oct-2-en-1-al. Furthermore, I do not believe that these are characteristic of bug colonies, which usually smell of decomposing bug faeces Nor, indeed, are these the only stinks one encounters in bug infested houses, which are often deficient in general cleanliness.

There is, too, a certain association between socially inadequate families and children chronically infested with head lice. So far as scabies and crab lice are concerned, it has always been held that their prevalence is encouraged by casual sexual encounters and low standards of hygiene. The recent resurgence of scabies (parallel to that in veneral disease) is very likely due to the neo-bohemian philosophy of many young people today; they combine a contempt for middle-class morality and for bourgeois cleanliness.

We have come to consider the hard-core residues of insect vermin as problems for 'the authorities'; and so they are. I shall discuss this aspect of official reaction to ectoparasites later. But there is little doubt that the best way of eradicating such pests is by universally improved standards of hygiene. Thus, body lice have virtually disappeared from the immense majority of citizens in civilized communities since people began regularly to wash their underwear. Human fleas are following and bed bug seem also to be getting rarer by virtue of improved housekeeping.

To maintain this progress, strong feelings if revulsion are no bad thing; but these must not get out of proportion. There are, however, a

certain number of people who have developed a neurosis about ectoparasites, which is sufficiently common to have been encountered by
advisory entomologists in several countries.[44,52,280] The condition,
which is variously described as 'entomophobia', 'acaraphobia',
'delusory parasitosis', '*wahnhafter Ungezieferbefall*' is comparatively
easy to recognize. The sufferer imagines that his (or her) body or house
to be infested with numerous minute biting parasites. This causes an
anguished feeling of being unclean and persecuted. Advice is sought
from chemists, doctors, advisory entomologists or (by post) from the
popular press. All sorts of preparations are tried, without avail and
some of the more vigorous measures can cause actual harm. One type
of sufferer characteristically brings 'specimens' to the laboratory, to
prove the existence of the infestation. These turn out to be pieces of
scurf or dirt, fibres or flecks from garments, small harmless insects, etc.
The delusions are not easy to recognize immediately, except by an
experienced entomologist, and not infrequently convince members of
the patient's own family. The sufferers are often people of intelligence,
who talk rationally except in regard to their unshakeable idea of being
infested. Two examples will make clear the type of problem involved.

Example (i) (from a man):

Even were the creature visible, it would still require ingenuity to catch one
without squeezing it to pulp.

The creature is not visible and to me it appears to travel as fast as an atom.

However, I will smear some glue on a microscope slide and hold it close to
my body in bed. With luck, one of the creatures may be held by the glue. . . . I
would say that the creature is smaller in size than the pores of the body. I often
feel their bite on the flat part of the heel. The bite is sharp and intense. When I
am still, I can feel their movements in my clothes. This gives rise to a most
horrid and fearsome sensation.

Example (ii) (from a woman):

The house I moved to was very old and should not have been sold . . . they
gave me some stuff to burn, no good.

Then I found a nasty black insect that got under the skin and worried me
nearly silly. I have tried everything, Lysol, kerosene, sulphur and carbolic acid.
I had a blister as large as a dinner plate on my chest and cannot get rid of the
horrid things in fact I have nearly killed myself. . . . I saw in the local Western
Argus that 200 children were not fit to go to school also I had some books from
the library full of nasty vermin, they were black. I wrote to the [a national
newspaper] and they said it was a kind of skin trouble *lots* of people had written

to them. They sent a 5% emulsion of Benzyl benzoate. I put some on my hair and find it affects my brain, I behave funny like I was drunk so I was afraid to use any more of it. . . . If you will advise me and let me know the *price* of your ointment I shall be *very very grateful*. I only wish I had never come to the dirty place (Kismet) the place should be burnt down. . . .

The job of an entomologist on encountering such people is to make quite sure that there is no real infestation and then to refer the case to a psychiatrist. Unfortunately, though the syndrome is fairly easily recognizable, neither the cause nor cure seem to be simple. Schrut and Waldron[280] suggest that repressed sexuality is often responsible but also mention other causes such as 'infantile or early guilt or other such effects, leading to rage, envy and aggression as well as other forbidden impulses'. This seems to include all the popular causes of neurosis.

Busch,[44] who quotes references as far back as 1894 ('Akaraphobie', described by a French dermatologist) gives as the first psychiatric reference one by a Dr H. Schwarz in 1929, who called it 'Circumskripte Hypochondrien' and placed it in the manic-depressive category. Elsewhere Busch refers to it as a paranoid phenomenon. He notes that the condition is chronic and may ameliorate in the course of years sometimes to spontaneous cure, but other cases remain as bad as ever.

Perhaps this digression into psychiatry may seem to have left the subject of the harmful effects of ectoparasites; so I will conclude with a connecting link. Very often, the trouble starts with an actual infestation, which leaves a deep impression on the sufferer, even when it has been eradicated.

4 The Prevalence of Ectoparasites in Different Epochs

Ancient and Classical Times

Ancient Egyptian writings, though mentioning 'respectable' insects like bees and ants, locusts and beetles, make no unequivocal reference to insect vermin. There are, according to Bodenheimer,[27] remedies for what may be lice and fleas in the Ebers papyrus, probably the oldest known medical compendium (1550 B.C.). Furthermore, the priests and scribes are always depicted as shaven, which according to Herodotus[128] (c. 480–408 B.C.) was so that 'no lice or any other impure thing should adhere to them while they are engaged in the service of the gods' (II. 37).

The Bible makes little reference to insect vermin (save, perhaps, to lice, which I discuss elsewhere). The flea mentioned in the Bible is rather more definite, though used in a metaphor, so that it does little more than testify to common usage of the word in a derogatory sense. About 1050 B.C., David, with perhaps a somewhat excessive humility, reproaches Saul for hounding him. 'After whom is the King of Israel come out? After whom dost thou pursue? After a dead dog, after a flea?' (1 Samuel 24: 14). Bed bugs do not warrant a mention in the Authorized Version. In Miles Coverdale's Bible of 1535, the 91st Psalm contains the slightly notorious phrase 'Thou shalt not nede to be afrayed for eny bugges by night'; but this is an older use of the word, before indeed the insects were known in England, corresponding to the word 'bogey'. References to scabies are doubtful, because of the difficulty of diagnosis. In Leviticus (13) any man suffering from a skin complaint was instructed to show it to the priest, who was expected to distinguish a leper from one merely suffering from the 'scab', who 'shall wash his clothes and be clean'. Then, among the afflictions with which the Lord was liable to smite the wrongdoer were 'the blotch', 'the scab' and 'the itch' (Deuteronomy 28: 27).

From classical Greece, we can find numerous references to ectoparasites for the first time. This is not because they were more verminous than other ancient peoples, but because they were highly literate, and interested in so many subjects. One such reference is in an early legend connected with Homer (*c.* 900 B.C.) which involves the ever popular theme of the triumph of the simple over the wise. It is the riddle of the fisher boys, who said: 'What we saw and caught, we left behind; but what we didn't see and failed to catch, we bring with us'. They were referring to head lice. This legend has a considerable history and has been traced back as far as Heracleitus[126] (sixth century B.C.). Another very ancient legend concerns Phalanthus, founder of Tarentum in the eighth century B.C. According to Pausanius[236] (second century A.D.), the oracle had promised him success, when rain should fall from a clear sky. In the early failures, his wife 'endeavoured to console him by her endearing officiousness; and as she was once supporting his head on her knees, and freeing it from vermin . . . began to weep; and her tears . . . fell on to the head of her husband, who then perceived the meaning of the oracle for his wife's name was Aethra, which is the Greek word for 'serene sky".' It is evident that, in these remote times, heroes and kings could be lousy. Another example was King Aegesilaus of Sparta (fifth to fourth century B.C.). While sacrificing to the gods, he was bitten by a louse. He killed it openly, remarking that his enemies would suffer likewise, even in the abode of the gods (Plutarch).[247]

In the *Sophist*, Plato[245] (428–348 B.C.), in the course of a discussion of similes, refers to the practice of searching for lice. 'In an argument', he said, 'one . . . does not in making comparisons think one more ridiculous than the other, and does not consider him who employs as his example of hunting, the art of generalship, any more dignified than him who employs the art of louse-catching, but only for the most part as more pretentious.'

Various more or less comic references to insect vermin turn up in the plays of Aristophanes[14]. In *Frogs,* Dionysus, enquiring about the journey to the Underworld, asks, 'And tell me too the havens, fountains, shops, roads, resting houses, stews, refreshment rooms, towns, lodgings, hostesses with whom were found the fewest bugs. . . .' In *Peace,* Aristophanes refers to the too frequent use by his rivals of jokes about such sordid subjects as rags and 'battles waged against Lice'. In *Clouds,* he pokes fun at the minute deliberations of philosophers, as follows:

. . . our great master Socrates to answer how many of his own lengths at a spring a flea can hop; for one by chance had skipped straight from the brow of Chaerophon to th'head of Socrates.

'And how did the sage Contrive to answer this?'

'Most dextrously. He dipped the insect's feet in melted wax which hardening into slippers as it cooled, by these computed he the questioned space.'

Kiel[168] discusses the extent of lousiness among the Greeks in classical times and concludes it could have been quite common. This would suggest that the great plague of Athens in the fifth century B.C. could have been typhus and he does not concur with Zinsser's[331] preference for smallpox, although admitting that the evidence is inconclusive. Educated Romans would, of course, be familiar with Greek legends. The dictator Sulla (138–78 B.C.) for example, quoted a Greek legend about a countryman and his lice to justify the killing of Q. Lucretius Ofella and to warn other refractory citizens. According to this legend, a farmer bothered by lice, stopped work twice to shake them out of his shirt. The third time, however, he decided upon drastic measures and burnt the shirt to avoid interruption in his work. 'And I tell you, who have felt my hand twice, to take warning lest the third time you need fire.' Appian[13] (The Loeb translation mentions fleas; but in the Greek version, the word seems to be φθειρες). In Roman literature, various forms of vermin turn up as metaphors and epithets. Catullus[56] (? 84–54 B.C.), for example, refers to bugs and spiders as little homely trifles:

Furi, cui neque servus est neque arca
Nec cimex neque araneus neque ignis

You, Furius, who possess neither slave nor money,
Nor hearth, nor even a bug or spider . . . (Poem 23)

Horace[148] (65–8 B.C.), on the other hand, uses the bug as a simile for an irritating back-biter: *Mem'moveat cimex Pantillus;* 'Do you think I'm moved by the bug Pantillus?' (*Satires* I. 10). Again, Plutarch (first century A.D.) compares flatterers to lice, which desert the body of a dead man.

Even in these ancient times, fleas were considered less disgusting than lice. Martial[199] (43–105) refers to a present of a back-scratcher to be used *pulice vel si quid pulice sordidus;* 'for a flea or anything worse' (XIV. 83). In another epigram, bed bugs are associated with poverty, for he derides an man for pretending to be merely poor when he is actually destitute, since he possesses *nec toga, nec focus est, nec tritus cimice lectus;* 'neither cloak nor hearth, not even a threadbare bug-infested bed' (XI. 32).

According to Friedman,[95] several Latin authors (including Horace,

Cicero, Ausonius and Juvenal) refer to scabies metaphorically, as something unsuccessful or unseemly: but he does not give references.

Late Mediaeval and Renaissance

After the decline of the Roman Empire, there stretches the long empty period of the dark ages, when little or nothing original was written. It is not until late mediaeval times that we can look again at secular literature in our quest. In the *Vision of Piers Plowman* by William Langland[178] (1330–1400), there is depicted a miser, *Avarida,*

> With a hode on his hed. A louse hatte above,
> And in a tawny tabarde of twelve wynter age;
> Al totorne and baudy and ful of lys crepynge;
> But if that a louse coulthe have lopen the bettre
> She sholde nogte haue walked on that welch so was it threadbare

Likewise, the *Canterbury Tales* by Chaucer[60] (1340–1400) the Friar tells of a fiiend in human form, who contemptuously remarks to the Summoner, 'A lousy jogeleur can deceyve thee'. Here, the epithet 'lousy', while derogatory, might well be apt. In the prologue to the Mancipal's Tale, there is a reference to vexatious fleas.

> What eyleth thee to slepe by the morwe?
> Hastow had fleen al night, or artow dronke
> Or hastow with some quene al night y-swonke
> So that thou mayst nat holden up thyn heed?

Fleas were, no doubt, particularly prevalent in the days when rushes took the place of carpets, even in lordly dwellings. These insects were observed to be attracted by fur-bordered garments, as the Knight of La Tour Landry[179] remarks in his book of instructions to his daughters (1370). In the fifthteenth century, too, the stole round a lady's shoulders was known as a flea-fur (*Flöhpelzchen*), because it was supposed to attract noxious insects.[28] Indeed, at various times up to the eighteenth century, some ladies wore special flea traps under clothing (see p. 185).

For some reason, women have traditionally been liable to attacks of fleas. Thomas Mouffet,[219] writing in the sixteenth century, notes that fleas 'are a vexation to all men but, as the wanton Poet hath it, especially to young Maidens, whose nimble fingers that are, as it were, clammy with moysture, they can scarce avoid'. The war between women and fleas was the subject of a comic German poem *Der Weiber Flöh Scharmützel,* to which I shall return later (see p. 116 and Fig. 4.1).

Fig. 4.1. Picture illustrating *Der Weiber Floh Scharmützel* a comic German poem probably by Johann Fischart (1545–1614). From Höllander *Die Karikatur und Satire in der Medizin*, Stuttgart (1905). By courtesy of The Wellcome Trustees.

At this time, not only were high born ladies liable to flea bites but even royalty was not immune. A story from the fifteenth century concerns Louis XI (1423–83). It appears that, one day, a louse crawled out of his finery into full view of the courtiers. One of his retinue, noticing the creature, quietly picked it off; but the King noticed his action and demanded what he had done. At length the man nervously confessed that he had found a louse on His Majesty. Instead of rebuking him, the King cheerfully improved the occasion by admitting that a louse is an excellent reminder to royalty that they are but human. Whereupon, a flatterer seized the opportunity to copy the servitor's example and pretended to find a flea on the king. But the latter (probably tiring of the subject) exclaimed, 'What! Do you take me for a dog, that I should be running with fleas? Get out of my sight!'

This story was versified in the following century by Johann Fischart[81] (1545–1614), a German satirist, under the title *Des Flohes Zank und Straus gegen die Stolze Laus* (see Fig. 4.2.).

Flöh Haß / Weiber Traß:

Der Wunder Vnrichtige /

vnd Spottwichtige Rechtshan=
del der Flöhe / mit den Weibern:
Weyland beschrieben

Durch

Zuldrich Elloposcleron.

Itzt aber von Newem abgestossen / behobelt /
gemehret vnd geziehret, mit vorgehendem
Lob der Mucken:

Vnd eingemischtem

Deß Flohes Strauß / mit der Lauß.

Alles kurtzweilig zulesen vnd wol zubelachen :
wo anders einen die Mucken nicht irre machen / oder
die Flöh einen plagen / die Läuß einen nagen /
vnd also von dem Lesen jagen.

Wer willkomb kommen will zu Hauß /
Kauff seim Weib diß Buch zum Borauß /
Dann hierinn sind sie Weg vnd Mittel /
Wie sie die Flöh auß Betthen schüttel.
Vnd hüt sich jedermänniglich
Bey der Flöh Vngnad / Biß vnd Stich /
Daß er diß Werck nit nach wöll machen /
Weil noch nit außgfürt sind die sachen.

Fig. 4.2. Title page of further German comic poems about fleas and lice. It includes the dialogue between the flea and the louse. The author, H.
Elloposcleros,[81] was a pseudonym for Johann Fischart, also known as Mentzer
(1545–1614). By courtesy of The Wellcome Trustees.

Hast du nicht die Geschicht gelesen
Wie einst ein Kaiser ist gewesen
Dem auf dem Kleid von ungefähr
Kroch eine biedre Laus daher,
Hin über die Achsel offenbar?
etc., etc.

which one may freely translate:

The Flea's trouble and strife with the proud Louse
Have you never read the story
How, on a King in all his glory
From underneath his robe, no doubt,
An honest louse by chance crept out,
And crawled across his back in view?
etc., etc.

By the sixteenth century, the Renaissance had reached northern Europe and there was a burgeoning of new literature, sufficiently modern in form to make an impact even today. In odd corners of this spate of poetry and prose, we find references to insect parasites. Thus, Thomas Phaire,[241] who wrote a *Regiment of Life* and a *Treatise on the Pestilence* is credited with the couplet:

Pediculus other whyles do byte me by the backe
Where fore dyvers times, I make theyr bones cracke.

Again, we find the profligate English poet, George Gascoigne[101] (1530–77) in a rather morbid mood;

The hungry fleas that frisk so fresh
To worms I can compare
Which greedily shall gnaw my flesh
And leave the bones bare.

In the *Tragical History of Dr Faustus* (1610) by Christopher Marlowe[198] (1564–93), there is a scene in which Wagner invites the clown to serve him, and mentions stavesacre (louse-bane). '*Clown:* Staues-aker? That's good to kill Vermine: then belike if I serue you I shall be lousy. *Wagner:* Why so thou shalt be whether thou dost it or no; for sirrah, if thou dost not presently bind they selfe to me for seuen years, I'll turn all the lice about thee into Familiars, and make them tare thee to peeces.'

It is only to be expected that Shakespeare[283] (1564–1616) with his enormous range of human intercourse, would include references to lice and fleas. We notice however that, as with Marlowe, these occur in dialogues of his minor or comic characters and are commonly used in epithets. There is a fairly strong implication that to be lousy is to be wretched and worthless. Thus:

> I care not to be the louse of a lazar.
> (*Troil. and Cres.* v.i)

> Wait like a lousy footboy at the chamber door.
> (*Henry VIII* v.viii)

> The codpiece that will house
> Before the Head has any,
> The head and he shall louse;
> So beggars marry many. (*Lear* III.ii)

At the same time, there is the beginning of a change from the literal to a figurative use. The combination 'lousy knave' appears frequently (*Merry Wives* III.iii twice; *Henry V* IV.viii and v.i-twice). Sometimes it is fairly evident that the meaning is more humorous than literal; as when Fluellen greets Pistol:

> God pless you Auncient Pistol
> You scurvy, lousy knave, God pless you! (*Henry V* v.i)

Fleas mentioned several times, often as similes for something tiny and contemptible rather than as pests.

> If a' has no more man's blood in his belly
> than will sup a flea. (*Love's Labours Lost* v.ii)

> He shall die the death of a flea (*Merry Wives* IV.ii)

> Thou flea, thou nit, thou winter cricket, thou!
> (*Taming of Shrew* IV.iii)

> If he were opened and you find as much blood as would
> clog the foot of a flea, I'll eat the rest of his anatomy.
> (*Twelfth Night* III.ii)

> A valiant flea that dares eat his breakfast
> on the lip of a lion. (*Henry V* II.iii)

Alternatively, fleas or lice can be subjects of whimsical jokes among the lesser characters. The 'boy' remarks of the recently deceased Falstaff:

> Do you not remember a' saw a flea stick on Bardolph's
> nose, and 'a said it was a black soul burning in hell fire.
> (*Henry V* II.iii)

As for puns, there is confusion between luce (an old name for pike) and lice. In the beginning of the *Merry Wives of Windsor,* Slender and Shallow are discussing the latter's pretensions to nobility, with Sir Hugh Evans:

Slen.: All his successors before him hath done't; and all his ancestors that come after him may; they may give the dozen white luces in their coat.
Shal.: It is an old coat.
Evans: The dozen white louses do become an old coat well;

There is also a mention of an itch mite, and the common practice of extracting it with a needle (see p. 204) in *Romeo and Juliet*(I.iv):

> Not half so big as a little round worm
> Pricked from the lazy finger of a maid

So far as bed bugs are concerned, I have already noted that they were scarcely known to Englishmen in Shakespeare's day. When the word 'bug' appears in his plays, it means a bugbear or bogey.

> Tush, tush, fear boys with bugs. (*Taming of the Shrew* I.ii)
>
> With O, such bugs and goblins in my life. (*Hamlet* v.ii)
>
> (Also in *Cymbeline* v.iii; *Winter's Tale* III.ii and *Henry VI* v.ii.)

the more prosaic attitude of an educated man in the sixteenth century may be summarized by a quotation from Mouffet's[219] *Theatrum Insectorum* (mostly written in the sixteenth century, though published later).

Fleas are not the lest plague, especially when in great numbers they molest men that are sleeping and they trouble weary or sick persons. . . . Though they trouble us much, they neither stink as Wall lice doe, nor is it any disgrace to a man to be troubled with them, as it is to be lowsie. They punish only sluggish people, for they will remove farre from cleanlie houses

The *Theatrum* says of the louse: 'It is a beastlie creature, and better known in Innes and Armies than is wellcome.' 'These filthy creatures, and that are hated more than dogs or vipers by our daintiest dames'.

The Seventeenth, Eighteenth and Nineteenth Centuries

Lice were still prevalent in high places, even in the seventeenth century, as shown by an early entry in the diary of Anne, Countess of Dorset,[127] dated 1603, when she was 13:

We all went to Tibbals to see the King, who used my aunt very graciously, but we all saw a great change between the fashion of the Court, as it was now and of that in the Queen's time, for we were all lousy by sitting in Sir Thomas Erskine's chambers.

Pepys,[238] also, remarks with irritation (18 July 1664):

Thence to Westminister, to my barbers, to have my Periwigg he lately made me cleansed of nits, which vexed me cruelly that he should have put such a thing into my hands.

And, later, (27 March 1667):

I did go to the Swan, and there sent for Jervas, my old Periwigg maker, and he did bring me a Periwigg, but it was full of nits, so I was troubled to see it (it being his old fault) and did send him to make it clean.

If Pepys was angry at the proximity of lice, there were other eminent seventeenth century gentlemen who might have welcomed the attention of at least one louse. The story has been often quoted, but I found it difficult to trace to its source. Cowan[68] (1865) quotes it from Southey's[292] Common Place Book (1849) in which one is referred to Pierre Bayle's[21] encyclopaedic *Dictionaire Historique* (1695–7). Herein is mentioned one Professor Blondel, who accompanied Louis-Henri Comote de Brienne on a grand tour from 1652 to 1655. A Latin account of the curiosities they had seen was published in 1662, and this contains our story, as a note by a M. Huet, on the mode of choosing a burgomaster in the town of Hardenburg, near Stockholm. It seems that those eligible sat round a table with the beards touching it and a louse was placed fairly in the middle. The man into whose beard the insect took refuge was then elected for the following year.

About this time Robert Hooke[143] was adding some rather fanciful remarks to his excellent drawing of a louse in *Micrographia* (1665).

This is a creature so officious, that 'twill be known to every one at one time or another, so busie and so impudent, that it will be intruding itself into every one's company, and so proud and aspiring withall that it fears not to trample on the best, and affects nothing so much as a Crown; feeds and lives very high and that makes it so saucy as to pull any one by the ears that comes into its way and will never be quiet till it has drawn blood: it is troubled at nothing so much as a man that scratches his head.

Over a hundred years later, the louse was still, apparently, 'affecting nothing so much as a Crown' according to an unsavoury incident at the court of George III. What happened is described by E. S. Turner.[306]

In 1787 there was a major domestic upheaval which began with the discovery of a louse on the King's plate at dinner. Its source was not clearly established, but George decided to take no chances and ordered all his kitchen staff to have their heads shaved. This decree was at once attacked as one more befitting an oriental despot. The cooks were proud of their hair and had no desire to walk out in the Mall looking like convicts. They addressed to the King a written protest which 'for boldness of language and the assumptions of importance it abounded in was scarcely ever equalled by any servant of any master in common life'. It appears to have made the reasonable point that the cooks were no more likely to breed vermin in their heads than any other Court servants. Nevertheless, all the cooks were shaved with the exception of one spirited youth who refused and was dismissed. The incident was not one on which fashionable ladies could afford to say too much, for their own elaborate head-dresses, when opened up after four or five weeks, were often found to be vermin-ridden.

Peter Pindar[242] (J. Walcot, 1738–1819) was not likely to overlook an affair like this; he published *The Louisad,* notable more perhaps for impudence than literary merit. The first of the four cantos begins:

> The Louse I sing, who from some head unknown,
> Yet born and educated near a Throne
> Dropped down (so will'd the dread decree of Fate)
> With legs wide sprawling on the Monarch's plate.
> . . . etc.

Fig. 4.3. The incident at the court of George III in which a louse was found in the monarch's dinner. From Pindar's poem about the incident, illustrated by Rowlandson.[111]

His satire was felt to merit prosecution, but the Lord Chancellor pointed out that the Crown could not be sure of a verdict; if the prosecution failed, they would all look a parcel of fools. The incident was illustrated by Rowlandson[111] (Fig. 4.3).

About the same time, Robert Burns[43] (1759–96) was expressing dismay 'on seeing a louse on a lady's bonnet at church'. He admits that:

> I wad na been surpris's to spy
> You on an auld wilfe's flannen toy;
> Or aiblins some bit duddie boy
> On's wyliecoat;
> But Miss's fine Lunardi! fie
> How daur ye do't?

Even in the early nineteenth century, it was still possible for high ladies to be, in the words of MacArthur,[190] 'twitted with their tendency to parasitical invasion'. He recounts the story of a quarrel of the third Countess of Holland with Theodore Hook (1788–1841). She finally turned him out of doors and said she didn't consider him worth 'three skips of a louse!' Hook retaliated with four lines of verse, as follows:

> Her ladyship said when I went to her house
> That she did not esteem me three skips of a louse;
> I freely forgive what the dear creature said
> For ladies will talk of what runs in their head.

In the last few pages, I have been largely concerned with the more intimate human parasites, which certainly appear in literature in many guises. One of the oddest is that of John Gay[104] (1685–1732) who invokes the same creature in a bizarre comparison:

> Brisk as a body louse she trips
> Clean as a penny drest.

Bed bugs were less often mentioned, partly because they were less familiar in northern Europe and partly, perhaps, because they seem to evoke little but aversion. During the seventeenth and eighteenth centuries, bed bugs became increasingly common; but only in the large cities. In 1682, Thomas Tryon writes of 'Vermin called Bugs that so many Hundred in this City, and other great Towns, are infested with; more especially in Holland, Italy, New England, Barbadoes, Jamaica and many other places.' John Southall[291] observes in 1730 that 'not one Sea Port in England is free; whereas, in inland towns, Buggs are hardly known.' He further adds,

Soon after the Fire of *London,* in some of the new built Houses, they were observed to appear, and were never noted to have seen in the old . . . Yet as they only seen in Firr-timbers, 'twas conjectured they were first brought to *England* in them; of which most of the New Houses were partly built, instead of the good Oake destroyed in the old.

In the latter part of the eighteenth century, parson Woodeford[325] came to accept bugs as a normal discomfort of visiting London. In May, 1782, he stayed at the 'Bell Savage' Inn on Ludgate Hill and he records that he was 'bit terribly by the Bugge last Night, but did not wake me.' This does not alter his opinion that, 'They were very civil people at the Bell Savage Inn, by name Barton and a very good house it is.'

Furthermore, he stopped there again for five nights on a subsequent visit to London, four years later, and once more suffered in the same way. This time, the bugs were so bad that, on the third night, he got up at 4 a.m. and walked about the city alone till breakfast. Of his last two nights, he says, 'I did not pull of my cloaths last night but sat up in a great Chair all night, with my Feet on the bed and slept very well considering and not pestered with buggs.'

The parson's predicament was well illustrated by two ribald pictures of Rowlandson[111] (1756–1827). *A Summer Night's Amusement* (1811) depicts two plump folk trying to collect and kill bed bugs (Fig. 4.4). On the wall is an advertisement for Mr. Tiffin, the bug eradicator to whom I have referred elsewhere (p. 82). A *Tit Bit for the Bugs* (1793) shows the dire effects of the bug bites (Fig. 4.5). A verse was appended to it as follows:

> Alas what avails all thy scrubbings and shrugs
> Thou hadst better return to thy sheets.
> Heap mountains of clothes over thee and thy bugs
> And smother the hive in the sheets.

In 1801, Jördans[158] remarks the bugs are not necessarily associated with poverty, since they are at least as common in the large mansions *(vielgadigen Pallästen)* of great towns as the small houses of little towns. A little later, the general outlook on insect vermin is sketched by Kirby and Spence in their textbook of entomology (1815). The louse is described as 'a very disgusting genus which Providence seems to have created to punish lack of attention to personal cleanliness.' Bugs are also deplored; but 'however horrible bugs may have been in the estimation of some, or nauseating in that of others, many of the good people of London seem to regard them with the greatest apathy, and take very lit-

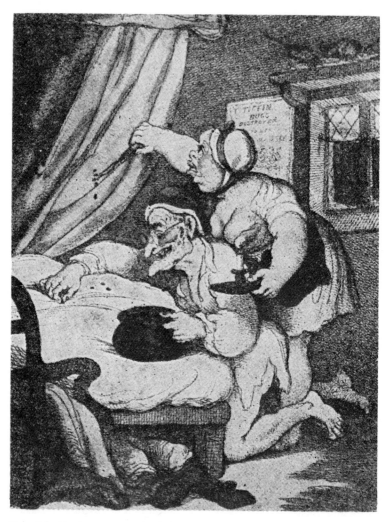

Fig. 4.4. 'A Summer Night's Amusement' by Rowlandson.[111] From an anonymous, pre-war commercial pamphlet. This print, though mentioned by Grego,[111] is not in the collections of the British Museum, the Victoria and Albert Museum or the Courtauld Institute.

Fig. 4.5. 'A Tit-bit for the Bugs' by Rowlandson, from Grego.[111]

tle pains to get rid of them.' Turton's edition of Linnaeus (1806) also refers to similar ubiquity of insect vermin. The bed bug is 'a troublesome and nauseous inhabitant of most houses in large Cities.' While the human flea 'inhabits Europe and America and is a well-known and troublesome insect in most families.' On the other hand, an article on Entomology in the Edinburgh Encyclopaedia remarks of the louse: 'On the continent of Europe, especially in Spain and Portugal, it is very abundant. In Britain, it is of very rare occurrence, and may have been introduced from the neighbouring countries.'

Despite this 'cleanlier than thou' attitude, bed bugs at least were still liable to concern the upper classes. Henry Mayhew[203] records a long and interesting conversation with a Mr Tiffin of Tiffin and Son, then described as 'Bug Destroyers to Her Majesty and the Royal Family'. An advertisement for this remarkable firm appears on the wall in the Rowlandson print (Fig. 4.4). The business was said to have been started about 1700 and was devoted to the upper classes; 'that is, for carriage company and those approaching it, you know'. Tiffin claimed to have 'noblemen's names, the finest in England' in his books, so we are prepared for the following intimate glimpse of royalty.

I was once at work on the Princess Charlotte's own bedstead. I was in the room and she asked me if I had found anything and I told her no; but just then I *did* happen to catch one, and she sprang up on the bed, and put her hand on my shoulder to look at it. She had been tormented by the creature, because I was ordered to come directly, and that was the only one I found. When the Princess saw it, she said, 'Oh, the nasty thing! That's what tormented me last night; don't let him escape!' I think he looked all the better for having tasted royal blood!

The royal blood in question was, I suppose, that of the unfortunate Charlotte Augusta, daughter of the future George IV, who died in child-birth in 1817, aged 23.

Modern Times

When I come to assess the prevalence of ectoparasites in Britain in the twentieth century, I should begin, perhaps, as an ordinary citizen, so that my experience could be contrasted with Samuel Pepys or Parson Woodeford. On this basis, it would seem that there has been a very great improvement in hygiene. I have never become accidentally infested with lice of any kind, nor have I ever slept in a bug infested bedroom. Twice I have acquired human fleas; but one occasion was 'in the line of duty' when I was investigating an infested air raid shelter during the War, so that this was rather a special case. The other occasion involved a visit to a cinema in Brighton, about 1936. Relaxing in an armchair on returning, I noticed a fine specimen feeding on my wrist. If it is a 'bold flea that takes its breakfast on the lip of lion', it is even bolder to feed openly on an entomologist. I groped for a specimen tube, captured the flea and his remains are now preserved on a microscope slide!

As it happens, I am a medical entomologist, so I might well be expected to have access to more extensive information on insect vermin than one of my literary sources of ancient times. As it turns out, however, the data are still fragmentary and often imprecise. I have sought the information from two sources. In the first place, public hygiene is the responsibility of local health departments, which publish annual reports. In many of these there are figures of the numbers of people treated for lice or scabies; and sometimes they give the numbers of houses disinfested from bugs or fleas. Unfortunately, the records are very variable especially when the staff changes. Some list 'pediculosis' under diseases of the skin and refer to de-lousing children euphemistically as 'cleaning' them. Those on which lice were discovered (during an often cursory examination) are recorded separately from those with nits only, as though the louse eggs were a different form of life. Houses with bugs or fleas are generally grouped together as 'verminous'. In fact, one gets the impression that ectoparasites are regrettable, but hardly worthy of serious attention, certainly beneath the notice of a qualified physician, so that they are relegated to health inspectors and nurses.

My second source of information is the experience gained in ten years as an advisory entomologist,[50] serving as a referee in the Public Health Laboratory Service (a duty still done by a colleague in my Department). The enquiries and specimens come from various sources: County and Borough Health Departments, Public Health Laboratories, Hospitals, commercial firms and miscellaneous (Ministries, Armed Forces, Physicians, Architects and the general public). Some of the most extensive information is available for scabies perhaps because, though not a notifiable disease, it is somewhat more interesting to medical men. Mellanby[211] collected data from hospital records in 1941 and seemed to show that there had been a steep rise up to that year, when about 2% of the entrants had scabies. Somewhat later in the War, in 1943, a record number of persons were treated for scabies in two large British cities (21 000 in Glasgow, 27 000 in Birmingham. These figures should not be taken to suggest that either city was more heavily infested than others in Britain, but rather that they kept some of the best records available to me.).[52] From this date, there was a fall in numbers (to 4000 and 7000 respectively) in 1947. By the 1950s, the figures had sunk to a few hundred; but from about 1960 on, there has been a steady rise; (the latest figures I have, for 1969, gave 3000 for Glasgow and 7000 for Birmingham). As I have said, these cities are not exceptional; Shrank and Alexander[285] (1967) point to a rise from 0.9% to 2.4% scabetic entries into a London hospital. Nor is this rise confined to Britain. A recent Czech paper points out that a similar rise has occurred in several European countries.[231]

The reason for the new scabies epidemic cannot be given with certainty. Some allege that high incidence produces widespread immunity in the population, so that incidence falls, until a new susceptibile generation comes along and begins the cycle again. I have never been convinced of this cyclical theory, because generations do not come in waves, but continuously. The other explanation, that of greater promiscuity and lower hygienic standards among modern young people, is supported by the concurrent rise in venereal disease.

So far as lice are concerned, there are notable contrasts in the fate of body lice and head lice. The former have become rare, except among vagrants who commonly sleep in their clothes. This progress towards rarity must have gained impetus during the last decades of the ninteenth century. Although there are no direct records, Buxton[54] used a graph to show the extinction of louse-borne typhus in England and Wales. From a figure of 4000 deaths p.a. in 1870, the graph falls sharply down to zero about 1918. Personally, I believe that this was due to reduction in lice, with improving standards of hygiene over the period. At present, the public health authorities mainly concerned with body lice are those at public assistance hostels and prison medical officers.

Head lice, however, constitute a different problem (as I have already indicated). Once again, I turn to Mellanby[210] for an early modern assessment. His first results, based on the years 1938–9 were very disturbing for he found that in some urban areas, nearly 50% of girls and over 30% of boys had head lice (based on hospital admissions). Later surveys showed little change by 1943, though there was definite improvement by 1947. After this, the decline has continued, but at a much slower pace, than one would hope. Recent figures show some 200 000 detected cases of head lice among schoolchildren and, as Maunder[202] points out, this figure could be quadrupled to take into account, firstly, the rather superficial examination on which it is based, and secondly, the incidence among babies and adults not included in the survey. He concluded that, adding figures for N. Ireland and Scotland, together with infestations by crab lice, it is probable that a million persons in the U.K. are harbouring these vermin on their persons.

Bed bugs are another story. Some of my earliest memories as medical entomologist concern the national drive to eradicate bugs from the slums in the 1930s. Committees were set up by the Ministry of Health and the Medical Research Council. Grants to study the problem were awarded, notably to my former colleague, C. G. Johnson[160] who produced voluminous publications on bed bug ecology. Little progress was made, however, until DDT appeared on the scene. During the War years, for example, bugs were taken on bedding bundles into a very large number of public air raid shelters. We did our best to check them with the smelly and only moderately efficient thiocyanate sprays. The introduction of DDT, however, entirely changed the situation. After this, bugs became a very minor problem. For example, in Glasgow, some 10.7% of houses were infested in 1934 and even in 1937 there were still 5.2%. From 1943 onwards, the figures fell, dipping below 1% between 1950 and 1953. In the last records I could find, they were at 0.01% in 1963. Similar progress has been noted in other countries (France, Italy, Denmark, U.S.A., U.S.S.R.).

Yet there is still some danger of a recrudescence. In many hot countries, bugs have become virtually immune to DDT, because of resistance. If some of these resistant bugs are introduced, in the luggage of some immigrant, they could spread and provide a serious control problem.

Finally we come to the human flea, *Pulex irritans,* which seems to be getting steadily rarer. My evidence for this statement, admittedly somewhat tenuous, consists of the declining numbers of complaints recieved by advisory entomologists,[31] most of which turn out to involve cat or dog fleas. Another equally fragile piece of evidence is the sorry plight of the last few flea circuses (see p. 123).

5   Human Reactions to Ectoparasites

Personal Hygiene and Propriety

Primitive peoples, like animals, have very little inhibition about dealing with ectoparasites; where they itch, they scratch. Furthermore, they catch and kill all they can. Occasionally, these activities have been recorded by artists. Thus, the *Little Beggar* of Murillo (1618–1682) (Fig. 5.1) is clearly searching for lice in the seams of his ragged shirt; while the typically candle-lit study by La Tour (1593–1652) (Fig. 5.2) is entitled *Woman with a Flea*. Incidentally, it is interesting to compare the serenity of this masterpiece with the frenzied caperings of the women in the comic German illustrations to Fishart and others (Figs. 4.1 and 4.2). The comic aspects of bug-hunting are also evident in Rowlandson's *Summer Night's Amusement* (Fig. 4.4).

So far as itch mites are concerned, it is just possible that Aristotle[15] was referring to them when he wrote of 'lice' which could be pricked out of little dry boils in the skin. Certainly the habit of pricking female mites out of their burrows has been adopted by infested people at various times and after extraction the mites were often burst between the finger-nails. This has been noted by various ancient writers, including the mediaeval Arabian scholars and early Renaissance writers in Europe (see p. 206).

Lice and fleas too are disposed of by bursting them between the finger-nails; but it is not unknown for them to be cracked between the teeth. This unhygienic habit, indeed, constitutes a serious risk in parts of the world where louse-borne relapsing fever is endemic. Nevertheless, it is by no means uncommon among primitive peoples, and when observed by divers imaginative travellers, it gave rise to stories of the use of lice as food. One of the oldest of these legends I traced back from Cowan[68] (1865) to Wanley[316] (1678), who cited Zwinger[332] (1571), who, in turn, referred to Sabellicus (1436–1506) 'ex Herodoto'. It concerns the Budini, a tribe of Scythians, who 'feed upon lice and the vermin that breed on the bodies of men.' I have not been able to trace this in

Fig. 5.1. 'The Little Beggar' by Murillo (1618–82), Paris, Louvre.

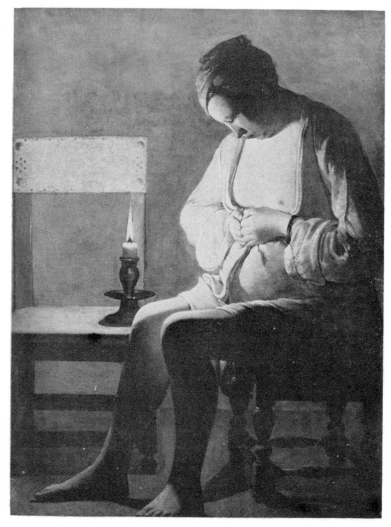

Fig. 5.2. 'Woman with a Flea' by La Tour (1596–1652), Musée Lorraine, Nancy.

Herodotus, who does however mention that the Libyans 'let their hair grow long and when they catch any vermin on their persons, bite it and throw it away' (IV.168).

The habit of chewing and swallowing lice was recorded in a 16th century Chinese manuscript, which describes some serious consequences; for the lice multiplied in the stomach of a man who had this habit, forming a mass which killed him.[139] Similarly, in the western hemisphere, Lionel Water,[313] exploring the Isthmus of Panama in the early eighteenth century, remarked: 'The natives here have lice in their heads, which they feel out with their fingers and eat as they catch them.'

It is not clear whether the lice were bitten to death for convenience or whether a kind of morbid satisfaction was involved. The most curious explanation was offered to Peter Kolben,[172] a German traveller who visited the Cape of Good Hope in the first half of the eighteenth century: 'They eat the largest of the lice with which they swarm; and if asked how they can devour such detestable vermin, they plead the law of retaliation and urge that it is no shame to eat those who would eat them.' This seems more of an excuse for justifiable pediculicide than an explanation of what we would regard as a disgusting habit.

The catching and killing of lice may call for a little friendly assistance. It is well known that gregarious monkeys indulge in 'grooming' and people watching this activity in the zoo generally believe that they are catching and eating each other's vermin. Zoologists tell us, in fact, that the morsels they capture and swallow are usually pieces of scurf; but I cannot help suspecting that any ectoparasites discovered would suffer the same fate. Indeed, the habit may have dual value in maintaining hygiene and social intimacy. Indeed, it seems very likely that certain habits of infested human beings are analagous. Mutual de-lousing may still be observed in unhygienic communities, on the lines illustrated in Fig. 5.3; and there are several examples in older literature. One of the earliest is in Purchas' Pilgrims (1625) concerning one Gonzalo Ferdinando de Orviedo, who wrote an account of the West Indies for Charles V of Spain. 'When these Indians are infected with this filthiness, they dresse and cleanse one another. And they that exercise this are, for the most part, women, who eat all they take and have such dexterity by reason of this exercise that our men cannot lightly attaine thereto.' More recently, a Mr A. R. Wallace[315] noticed the same habit in South America, and reported it to the Entomological Society of London, as follows: 'This Apterous insect, which is eaten by the South American

Indians, more I presume as a delicacy than as an article of food, is a species of *Pediculus,* which inhabits the heads of that variety of mankind, and is probably a distinct species from that of our own country ... A couple of Indian belles will often devote a spare half hour to entomological researches in each other's tresses, every capture being immediately transferred with much gusto to the mouth of the operator.'

Fig. 5.3. 'The Madras Hunt a primitive, wasteful of time and unsatisfactory method of delousing'. After Patton, *Insects, Ticks, Mites and Venemous Animals*, Croydon (1931).

His account is redolent of the ponderous, arch, humour of the ninteenth century, with its 'Indian Belles' and 'entomological researches'. A curious variation of the habit was recorded by A. Steadman,[294] in his *Wanderings and Adventures in the Interior of South Africa* (1835).

It often happens that one caffir performs for another the kind office of collecting these insects, in which case he preserves the entomological specimens, carefully delivering them to the person to whom they originally appertained, supposing according to their theory that, as they drew their support from the blood of one man, should they be killed by another, the blood of his neighbour would be in his possession, thus placing in his hands the power of super-human influence.

I suppose it is likely that our European ancestors practised 'phthiropophagy'; certainly there is no doubt about the habits of delousing one's family and friends, which has been recorded by several well-known artists. Usually the subject seems to be a mother delousing a child, as in the picture by Murillo (Fig. 5.4) and de Hooch (Fig. 5.5);

Fig. 5.4. 'The Toilet' by Murillo (1618–82), Munich, Alte Pinakothek.

Fig. 5.5. 'Mother Delousing Child' by P. de Hooch. Rijksmuseum, Amsterdam.

but in the two variations of the school of van Ostade, the 'patient' is a lubberly oaf, though the old woman de-lousing him may still be his mother (Fig. 5.6).

Another type of affectionate link exemplified by human grooming is that between husband and wife or sweetheart. I have already mentioned the 'endearing officiousness' of the wife of the Greek hero Phalanthus, 'supporting his head on her knees, while freeing it from vermin'.[236] Leaping forward 24 centuries to the *Pastoral Dialogue* of Jonathan Swift[297] (1667–1745), we find two bucolic Irish sweethearts after a tiff.

> *Dermot:* When you saw *Tady* at long bullets play
> You sat and lous'd him all the sunshine day.
> How could you *Shulah,* listen to his tales,
> Or crack such lice as his between your nails?
> *Shulah:* *Dermot,* I swear, tho' *Tady's* locks could hold
> Ten thousand lice, and every louse of gold
> Him on my lap, you never more would see;
> Or may I lose my weeding knife—and thee.

Fig. 5.6. Delousing in two versions of a picture (?) by Adrian van Ostade. Collection Hanfstaengl, Munich.

From the affectionate service of grooming to remove vermin, it is a short step to the somewhat more intimate relieving of the irritation. The tipsy Stephano in *The Tempest* (ii.ii) sings of a virago of easy virtue, who was unfair to sailors, in that

> She lov'd not the savour of tar or of pitch
> Yet a tailor could scratch her where'er she did itch.

Fig. 5.7 from the early nineteenth century *Album Comique de Pathologie Pittoresque* shows a row of military types of queueing to offer the same service to a young lady afflicted with scabies.

Fig. 5.7. 'The Itch' by von Bellange from Höllander, *Die Karikatur und Satire in der Medizin*, Stuttgart (1905). By courtesy of The Wellcome Trustees.

We have been descending, somewhat, to low life, however. Even in ancient times, it was generally recognized among cultured peoples that killing vermin in public was not respectable. I have mentioned King Aegesilaus of Sparta (fifth to fourth century B.C.) killing a louse while sacrificing to the gods; but he felt it necessary to offer a kind of defiant apology (p. 68). In the Babylonian Talmud,[298] we have '... our Rabbis taught one must not search (his garments) in the street, out of decency ...' (*Shabbath* 12a). Similarly, the Moslem 'Hadith' tradition recorded by Damiri forbade the killing of lice in holy places, especially in the Mosque (Bodenheimer).[27]

As to manners in Europe during the Dark Ages, we are largely ignorant; but by the fifteenth century, there are indications that such behaviour was frowned upon in the best circles. About 1450, John Russell,[273] who describes himself as 'Sum tyme Seruande with Duke Umfrey of Glowcetur, a Prynce Fulle Royalle, with whom Vshere in Chambur was Y, and Merchalle also in Halle' wrote a book of instructions for the superior servants of the nobility. This *Boke of Nurture followyying Englondis Gise* contains the following:

> Simple condicyons of a person that is not taught
> I wille ye eshew, for euermore they be nowght,
> Your hed no bak ye claw, a fleigh as though you sought,
> Ne your heer ye stryke, ne pyke, to pralle for a fleshe mought.
> (Don't scratch your head or back as if after a flea;
> Or your hair as though seeking a flesh mite—? a louse).

A somewhat similar code of manners was compiled rather later on the Continent, at the Jesuit Seminary of La Fleche. In 1595, the pensionaries of this college sent to those of the college at Pont à Mousson a treatise entitled *Bienseance de la Conversation entre les Hommes*. This achieved great success. By order of the Bishop of Tours, it was translated into Latin and subsequent editions appeared in Spanish, German and Bohemian. An English translation appeared in 1640, purporting to be the work of an eight-year old boy, one Francis Hawkins. The French and English versions of Rule 22 begin thus:

Gardez vous bien de vous arrester a tuër une puce, ou quelque sale bestiole de cette espece, en presence de qui que a puisse estre . . .
Kill not a flea or other uncleane vermin in the presence of others . . .

Zinsser[331] quotes one Reboux, writing on the education of a princess of France in the seventeenth century, as follows:

One had carefully taught the young princess that it was bad manners to scratch when one did it from habit and not by necessity, and that it was improper to take lice, fleas or other vermin by the neck to kill them in company, except in the most intimate circles.

I cannot trace this reference, but it seems possible that it was concerned with the code of rules originating at La Fleche. What is, however, certain, and of considerable interest, is that the English version of these rules was copied out by George Washington, when a boy. The links in this curious chain were discovered by D. M. Conway[66] in 1890. He points out that Washington went to school at Fredericksburg, where he was probably taught by a French Huguenot emigre called Mayre, born in Rouen at the end of the seventeenth century. Mayre, who lived for some years in England, might have been familiar with both the English and French versions of the Rules.

In the last hundred years, the advances in hygiene have rendered it unnecessary to include reference to de-lousing in books of etiquette. No doubt it is still considered ill-mannered to scratch oneself, though head scratching is normally associated with cogitation rather than infestation. However, the very idea of creeping vermin can induce itching sensations in many people, as witness the following story.

During the Second World War, when I was working on remedies for body lice, we were somewhat concerned about the possible spread of such parasites in air-raid shelters. As a precautionary, I was asked to give some simple lectures on the subject to groups of London shelter-marshals; and afterwards, the departmental secretary dispensed appropriate leaflets. It appears that one East-end shelter-marshal remarked to her 'Fair made me itch all over, it did, Miss; and I noticed the young gentleman was scratching 'is 'ead while 'e was talking!'

Insects and Hygiene

In most ancient civilizations, the more cultured and affluent individuals probably kept themselves free from insect parasites by cleanliness. I have already mentioned the Egyptian priests who regularly shaved their bodies to this end. This hygienic behaviour was, no doubt, empirical in origin. In the long period of Aristotleian biology, it was reconciled with the master's teaching by the necessity of dirt or excretions for the generation of insect vermin. There is, however, a hint of a more modern outlook in the injunction of the Babylonian Talmud[298]

Vermin in linen; if one launders his garment and does not wait eight days before putting it on, the vermin, which may still be in it, are produced and harmful. *(Pesahim,* 112b)

Bartholomew de Glanville (or Bartholomew Anglicus)[19] writing about 1240, says that 'against the grieving of lice, oft washing, combing and medicinal cleansing of the head helpeth . . .' And he notes that 'A sluttish kept house breedeth fleas . . .'

Cleanliness is advocated as a defence against lice in various editions of the fifteenth century German *Hortus Sanitatis*.[254] An English 'pirated' translation under the title *The Noble Lyfe and Nature of Man*,[149] in the early sixteenth century, includes the following:

A louse is a worm with many fete and it cometh out of the filthi and onclene skynne . . . To Withdryue them, the best is to washe oftentymes and to change oftentymes clene lynen.

The *Insectorum Theatrum*[219] attributes a comparative freedom from lice to the superior cleanliness of Englishmen.

As for dressing the body; all Ireland is noted to this, that it swarms almost with Lice. But that this proceeds from the beastliness of the people and want of cleanly women to wash them is manifest, because the English that are more careful to dress themselves, changing and washing their shirts often, have escaped that Plague. Hence it is that armies and prisons are so full of lice, the sweat being corrupted by wearing always the same clothes, and from this arises matter for their origin, by the mediation of heat.

Elsewhere, Mouffet makes similar suggestions about bed bugs. Speaking of the suggestion that the abstention of the Carthusian monks from meat renders them less liable to bugs, he says:

. . . he should rather have alleged their cleanliness and the frequent washing of their beds and blankets to be the cause of it, which when the *French*, the *Dutch* and *Italians* do less regard, they more breed this plague. But the *English* that take care to be cleanly and decent are seldom troubled with them.

(John Ray[261] (1710, posth.) refers to these remarks and more modestly ascribes the difference to the Continental climate, the violent heat of which he considers to be more favourable for the propagation of bed bugs.)

The Italian Aldrovandi[6] (of whom more later, p. 137) discusses the relations between dirty conditions and bed bug infestations in his book *De Animalibus Insectes* (1603). 'They infest both the chambers of rich and poor, but are much more troublesome to the poor. The reasons for this relate to the material from which they breed and the cleaning which prevents their breeding. For they do not breed in beds of which the linen and straw is frequently changed, as in the houses of the rich. Because they arise from human dirt, or, as Aristotle says, from humours from the animal body. And these tend to accumulate in beds of the poor, where straw and linen are rarely changed' (VI.2).

The theory that vermin are generated from 'the excrements and

Breathings of the Body' and that they can be avoided by cleanliness and fresh air is repeated by Thomas Tryon[304] in his book. *A Treatise of Meats and Drinks . . . the Excellency of Good Airs, and the Benefits of Clean Sweet Beds. Also of the Generation of Bugs and their Cure.* (1682). Cleanliness was also supported by those who abandoned the idea of spontaneous generation. Johannes Sperling[293] (1661), for example, repeats Aldrovandi's words, but insists that bugs breed *in* dirt not *from* it (*sed in excrementis potus discendum, quam ex excrementis*). Leeuwenhoek[181] (1696) also notes the difficulties the poor suffer with lice: 'This animal which is so troublesome to many, especially to the poor who have not the means of frequently changing their apparel . . .'

Southall[291] (1730), who was concerned to extol his own insecticide for the eradication of bugs, nevertheless points out the advantages of beds being demountable and easy to clean. Indeed, he undertakes (for a suitable fee) to alter bedsteads and make them easy to take apart and reassemble. De Geer[105] (1773) and Joerdans[158] (1801) both appear to be surfeited by the lists of insecticides given by ancient and contemporary writers. They advocate thorough cleansing for the eradication of bugs, followed by re-plastering of the walls to seal up all harbourages of the insect. Until the introduction of such potent modern insecticides, such as DDT, this advice could scarcely be bettered.

Vermin and Virtue

The intellectuals and idealists, however, were not always supporters of hygiene. Thus, an ascetic neglect of personal comfort (and, incidentally, hygiene) was a tenet of that prickly exhibitionist, Diogenes (412–323 B.C.) and his followers, the Cynics. To some extent, this was a reaction against the luxury and indulgence of the rich and powerful; but one cannot help suspecting that the enjoyable shocking of the smug was an element of the Cynics' flouting of propriety.

Their ideals were revived by that singular emperor, Julian the Apostate[162] (331–63). In a book directed against the effeminacy and corruption of the inhabitants of Antioch, he describes. with ironic satisfaction, his own unkempt appearance and his shaggy beard 'with the lice that scamper about in it, as though it were a thicket of wild beasts.'

This excessively ascetic attitude was enthusiastically adopted by the early Christian hermits, of whom D'Israeli[75] remarks,

These monks imagined that holiness was often proportional to a saint's filthiness. St Ignatius, it was said delighted to appear abroad in old dirty shoes; he never used a comb but let his hair clot; and religiously abstained from paring his nails. One saint attained such piety as to have 300 patches on his breeches; which, after his death, were hung up in public as an incentive to imitation.

Bertrand Russell[272] remarked on the same subject,

Cleanliness was viewed with abhorrence. Lice we called 'pearls of God' and were a mark of saintliness. Saints, male and female, would boast that water had never touched their feet except when they had to cross rivers.

This attitude certainly persisted in the twelfth century, for it is recorded that Thomas à Becket was extremely lousy at his decease. MacArthur,[190] who inspected the original manuscript in Canterbury Cathedral, writes as follows:

Thomas à Becket was murdered in Canterbury Cathedral on the evening of 29 December (4 January by our calendar). The body lay in the Cathedral all night, and the next day, after some debate, it was decided to remove the clothing in preparation for the burial. The dead Archbishop was clad in an extraordinary accumulation of garments. Outermost there was a brown mantle; next a white surplice, underneath this, a fur coat of lamb's wool, then woollen pelisse; then another woollen pelisse; below this, the blackcowled robe of the Benedictine order; then, a shirt, and finally, next to the body, a tight fitting suit of coarse hair cloth, covered on the outside with linen, the first of its kind seen in England. The innumerable vermin which had infested the dead prelate were stimulated to such activity by the cold that his hair cloth garment, in the words of the chronicler, 'boiled over with them like water simmering in a cauldron', and the onlookers 'burst into alternate fits of weeping and laughter, between the sorrow of having lost such a head and the joy of having found such a saint'.

In the meantime, another reason for not destroying insect vermin was found in an extreme concern for the sanctity of life. In about the fifth century A.D. the compilers of the Babylonian Talmud[298] discussed the things which are lawful on the Sabbath: 'Rabbi Eliezer said: "He who kills vermin on the Sabbath is as though he killed a camel on the Sabbath!"' (*Shabbath*, 107b). This leads to a rather hair-splitting argument as to whether lice can indeed be regarded as valid forms of life, in the eye of the Law. Elsewhere, there is a discussion as to whether the searching of garments in the street is permissible, which ends '... our Rabbis taught: if one searches his garments (on the Sabbath), he may press (the vermin) and throw it away, providing he does not kill it' (*Shabbath*, 12a).

According to Purchas[255] (1625), on the Jewish Sabbath, 'The Lampes must not be put out nor the light, thereof applied to the killing of fleas, to reading or writing, etc.' (II.17). Here, however, the emphasis is on observing the Sabbath rather than avoiding flea murder.

Excessive religious zeal has led the more fanatical in various lands to avoid killing lice and fleas. This caused some surprise to early travellers in the Orient. Thus, Marco Polo[197] (1254 – 1324) tells of the Chughi (Yogi) of the kingdom of Lar 'of whose abstinence and hard strict life I will tell you . . . I will adds too, that for nothing in the world would they kill a living creature or any animal—be it fly or a flea or a louse or any kind of vermin for they say they have souls.'

One cannot, however, take such benevolence quite seriously; and indeed, the sparing and cherishing of vermin does not necessarily guarantee virtue in other matters. Thus, in Purchas' Pilgrims[255] (1625) we read:

If lice do much among the natives of Calabar and Malabar, they call to them certain Religious and holy men after their account; and these Observants will take upon them all those lice which the others can find and put them on their heads, there to nourish them. But yet for all this lousie scruple, they stick not to cozenage by false weights, measures and coyne, nor at usury and lies.

Somewhat later certain travellers to Surat noticed some curious proceedings at a hospital for sick animals. Not only were the usual kinds of animals nursed back to health, but even pests and vermin were cherished. According to the Churchills' *Collection of Voyages and Travels* (1732)[63] one Gemelli visited this hospital in 1695 and saw: 'a poor wretch, naked, bound down hands and feet, to feed the Bugs or Punaises brought out of the stinking holes for that purpose.'

About 80 years later, James Forbes,[93] who spent some 30 years in India, visited the same Banian Hospital. In his *Oriental Memoires* (1813) he wrote

. . . the most extraordinary ward was that appropriated to rats, mice, bugs and other noxious vermin. The overseers of the Hospital frequently hire beggars from the streets for a stipulated sum, to pass the night amongst fleas, lice and bugs on the express condition of suffering them to enjoy their feast without molestation.

In the meantime, the philosophy of the sanctity of animal life had invaded the Christian religion. Most of us are familiar with the reputation of Francis of Assisi (1182–1226), though we tend to associate the benevolence with birds and mammals. However, it was St Francis who is credited with calling lice the 'pearls of poverty'. And Purchas,[255] in a footnote to his story of the oriental holy men who freely accepted lice from others, says robustly, 'The like lousie trick is reported in the legend of St Francis and in the life of Ignatius, one of the Jesuitical pillars!'

Somewhat later than St Francis lived Cardinal Bellarmine (1542–1621), a bastion of the counter-reformation who, according to Lehane,[182] displayed his saintly altruism in the matter of flea bites. 'We shall have Heaven to reward us for our sufferings', he said, 'but these poor creatures have nothing but the enjoyment of this present Life.' Lehane further points out that Bellarmine's pupil turns up in the Ingoldsby Legends as St Aloys, Bishop of Blois.

> He grieved and pined
> For the woes of mankind
> And of brutes in their degree
> He would rescue the rat
> From the claws of the cat
> And set the poor captive free;
> Though his cassock was swarming
> With all sorts of vermin
> He'd not take the life of a flea!

Another of this ilk was St Macaire who, according to D'Israeli[75] '. . . was so shocked at having killed a louse that he endured seven years of penitence among the thorns and briers of a forest.'

Moliere seems to have this in mind in describing the character of Tartuffe. 'He considered as sins the merest trifles to the point when, the other day, he caught a flea while praying and reproached himself for killing it with undue spite.'

The tradition of saintliness and self-neglect persisted in certain religious circles until comparatively modern times. Henry Meige,[209] writing in 1897, remarks, 'In our own century lice reach their apotheosis! In 1873 a papal decree proclaimed the canonization of a wretch who gained heaven clothed in rags and being eaten by vermin; with the Blessed Joseph Zabre, lice have been sanctified.'

In the meantime, however, a totally different view had been propounded by John Wesley (1703–91): 'Let it be understood that Slovenliness is no part of religion; that neither this nor any text of Scripture condemns neatness of apparel. Certainly, this is a duty not a sin. Cleanliness is, indeed, next to godliness' (*Sermon, 93*). The resounding final phrase (which may be derived from a Hebrew tenet) struck a responsive chord in nineteenth century England. Today, however, 'neatness of apparel' has become identified with the 'Establishment' among rebellious youth. The modern hippies have perhaps something in common with the original Cynics who similarly rebelled against convention and were likened to dogs (Gr. κυνικός) because of their shameless and agressive manners.

Ectoparasites Considered as Divine Afflictions and as Morbid Curiosities

Certain folklore tales about insect vermin regard them as a kind of scourge to mankind; for example, a legend of the Sandwich Islands recorded by J. S. Jenkins[157] in 1852 is strongly reminiscent of the tale of Pandora's box. It concerns a woman of Waimea who left the island to meet a lover. When she returned, he gave her a present of a bottle, which she was on no account to open until her return. When she finally opened it, thousands of fleas escaped, which have been hopping and biting ever since. Cowan[68] mentions another legend formerly current among the Kurds of Eastern Turkey. It appears that Noah's Ark sprang a leak, which the serpent offered to seal, if he could be granted a meal of human flesh after the Deluge. Noah reluctantly agreed and the serpent drove his body into the hole and stopped the leak. Later he demanded fulfilment of his pledge. Noah, on the advice of Gabriel, burnt the document and scattered the ashes, which flew about the world, turning into Fleas, Flies and Lice, which have tormented us ever since. In Southey's[292] *Common Place Book* is the story of the Devil teasing St Dominigo in the shape of a flea. The Devil rashly leapt onto the saint's book, whereupon the holy man closed it with a bang and used him as a bookmarker for the rest of the volume. Sometimes ectoparasites have been regarded as a serious retribution. Most people will remember that lice are mentioned in Exodus (8:16) as one of the plagues visited on the Egyptians, about 1500 B.C. 'And the Lord said unto Moses, Say unto Aron, stretch out thy rod and smite the dust of the land, that it may become lice throughout all the land of Egypt. And they did so; for Aaron stretched out his hand with his rod and smote the dust of the earth, and it became lice in man and in beast . . .' (Authorized Version). The pests are described by the Hebrew word *Kinnim* or *Chinnim*, which is rather ambiguous, having the sense of 'insect vermin', so that one cannot be quite certain that lice were meant.

Scabies may have been mentioned in the Bible as another scourge; but its differentiation from other skin conditions is by no means easy. In Deuteronomy (28:27) about 1400 B.C., Moses warns the Israelites of the wrath of God if they transgress. Among other dire punishments 'The Lord will smite thee with the blotch of Egypt and with the emerods and with the scab and with the itch, whereof thou canst not be healed.'

In ancient Greece, there began a tradition about lice even more sensational and obscure that the plague of Egypt. This was *phthiriasis,* or the lousy disease; the *'morbus pediculosis'* of mediaeval scholars, in which the sufferers were supposed to generate large quantities of lice from their skins, which ate away their flesh until they perished miserably. Accounts of this 'disease' recur through the centuries and, curiously enough, similar cases have been found in mediaeval Chinese medical literature. In the West, the stories persist through the Middle Ages and the Renaissance up to the early nineteenth century. It is a curious and puzzling matter and I will discuss possible scientific explanations in a later chapter. At this stage, it is worth noting that he people said to have been affected in classical times, were by no means filthy outcasts of society, but often eminent men. As Knott[171] points out, any of them had made themselves decidely unpopular with large numbers of their contemporaries, so that accounts of their degrading afflictions would have been circulated with relish (rather like the tales of Adolph Hitler crawling on the floor and biting the carpet). The names were collated by mediaeval encyclopaedists (notably Theodore Zwinger[332] of Basel) together with a brief mention of their misdeeds, so that the lousy disease often seems a divine retribution. The list of victims includes the following: Cassandros (355–297 B.C.) a wicked king who slew the mother and sons of Alexander; Quintus Pleminius (*c.* 200 B.C.) who despoiled the Temple of Proserpine in Rome and was imprisoned for it; the dictator Sulla (138–78 B.C.), not only a violent but a dissolute man; the emperor Flavius Cladius Julianarius the Apostate (*c.*330 A.D.) who desecrated the high altar in Antioch by urinating on it; Honoricus, King of the Vandals (384–423) who exiled 444 Christian bishops; Radbertus a Saxon (*c.* 650–80 A.D.) who killed a bishop in Auvergne; Arnulphus, lord of Normandy in the ninth century, who displayed excessive and intolerable pride; Leostanus (*c.* 855) who doubted the miraculous growth of hair and nails of the dead English king Edmund and demanded the body to be shown to him (When brought face to face with it, he went mad and soon after died of phthiriasis); Fulcherus, bishop of Noyon, in Flanders (855) whose body seethed with lice and who was buried sewn up in a deerskin. This was said to be a notable example of divine punishment for bribery and simony; Pipinus (Pepin) (1017) reprimanded by the holy Lambertus for taking the wife of Alpiadis as a concubine and killing Alpiadis.

The punishment theory, however, will not account for all the relish evident in the accounts of the lousy disease. One cannot avoid the impression that there exists a morbid fascination with this gruesome subject, which will be evident when we read later and more detailed accounts. But, before we consider these, we should take note of a very curious traveller's tale by Diodorus of Sicily[76] III.29 (fl. first century B.C.). In one of his extensive books on foreign parts he describes the North African tribe of Acridophagi or locust eaters.

A short distance, on the edge of the desert, dwell the acridophagi, men who are smaller than the rest, lean of body and exceedingly dark ... in the spring season, strong west and south-west winds drive out of the desert a multitude of locusts ... From these locusts, they have food in abundance all their life long ... As for the manner in which they end their lives, not only is it astounding, but extremely pitiful. For when old age draws near [in fact, he says they seldom live older than 40] they breed in their bodies winged lice, which not only have unusual form, but are also savage and altogether loathsome in aspect. The affliction begins in the belly and the breast and, in a short time, spreads over the whole body. And the person so afflicted, is at first irritated by a kind of itching and insists on scratching himself ... but as his hands tear at his body, such a multitude of vermin pour forth, that those who try to pick them off, accomplish nothing, since they issue forth one after another, as from a vessel that is pierced throughout with holes. And so these wretches end their lives in a dissolution of the body in this manner ... either by reason of the peculiar character of their food or because of the climate.

I suppose that Plutarch[248] (46–120 A.D.) may have read this account and it is possible that it may have influenced his description of the death of Sulla a hundred years before (though he does not, indeed, refer to the fate of the acridophagi).

... his whole body became one mass of lice (*corpus in pediculos totum versum*); and though many persons were employed day and night removing the lice, yet they were unable to destroy as many as were produced, so that his clothes, bath, furniture, wash hand basins and food were full of them. And though he bathed frequently, every day and washed and rubbed his body, yet this was of no avail. For the transformation of his body into lice was so rapid that all attempts at cleansing were frustrated. (*Lives* IV.439)

This somewhat fantastic exaggeration is equalled by an account of a nobleman of Tabora in Lisbon described by a famous sixthteenth century Portuguese doctor, Giovanni Roderiguez de Castillo Bianco, generally known as Amatus Lusitanus[10] (1556). Of this man, Amatus declares, his body so abounded in lice that two of his Ethiopian slaves were employed in emptying baskets of them into the sea. Again, Phillip II of Spain, who died in Escurial, Madrid, in 1597, suffered for a long time from a severe abscess of the right knee, and after this was opened, four others formed on his chest; and while four persons held him suspended in linen, two others by turn could scarce sweep away the lice . . .

Marcellus Donatus,[196] an Italian physician who practised in Venice, gives in *Historia Medica Mirabili* (1613) a fairly full list of the historical cases of the lousy disease and adds one of his own. It concerns '. . . a person of high rank, extremely fat, whose belly was eroded and mortified by little worms, engendered in his skin, which was excessively distended by fat and humours; and these worms were not unlike those produced in old rotten cheese.' This strikingly resembles the affliction of a Lady Penruddock of whom Mouffet (1634) says: 'acari swarmed in every part of her body—her head, eyes, nose, lips, gums, the soles of her feet, etc., tormenting her day and night, till in spite of every remedy, all the flesh of her body being consumed, she was at length relieved by death.' We note that mites or 'acari' are mentioned here, a matter of which I will consider further later (p. 208). Later in the seventeenth century, various peculiar cases were cited in the records of the German Scientific Academy of Leopoldina. Publications of this society deal with various bizarre curiosities, such as a young woman with three nipples on each breast (illustrated) and another with a hair-lined vulva!

Two of the articles[235] published by the Academy in 1687 specifically mention lice. One (Art. 42) refers to a man whose '*Penis cum scrote a pediculus corrosus et consumptus*' The other (Art. 60) concerns a country girl who, to cure jaundice (see p. 180) on her mother's advice, swallowed some live lice taken from her own hair. As a result, the creatures multiplied inside her, giving rise to a form of phthiriasis, from which, however, she was cured.

A third article (Art. 38) recalls the cases of Lady Penruddock and the person of high rank mentioned by M. Donatus. It reports the case of a Frenchman whose blood was so corrupted that very minute animals came forth day and night, with horrid tortures, through most of the outlets of his skin, as the eyes, nose, mouth and bladder; and at length put an end to his miserable life.

At this point, I propose to leave to leave for the present the strange story of the 'lousy evil'. It is not possible, of course, to assign historical changes to an exact point in time; but, somewhere around the year 1600 a more modern outlook began to appear. In the present story, the fabulous and bizarre accounts of the late Middle Ages gradually gave place to more plausible and detailed records. It is true that some of these are probably erroneous and others inexplicable; but it seems reasonable to consider them in a history of the medical aspects of ectoparasites, rather than in the present chapter.

Fleas and Moral Precepts

Talking animals in stories or verses which point a moral, have been popular since the far-off times of Aesop (619–564 B.C.), whose well known Fables were handed down verbally and then transcribed many times. Insects such as ants, bees and grasshoppers, turn up in some of these tales and one might also expect to find creatures as familiar as insect vermin. As it happens, fleas are protagonists in fables of Aesop, La Fontaine and later writers; but bugs and lice do not seem to be suitable.

Fleas occur in at least two of Aesop's[3] fables. In one of these, a flea steals a ride on a heavily laden camel, but later jumps down to relieve the beast. The camel thanks him for nothing, the moral being (in the words of Caxton's 1482 version) 'and therefore of hym which may neyther helpe ne lette men, nede not make grete estymacion of.'

The other tale concerns a man who caught a flea, asking 'why bytest thow me?' The flea pleaded that, 'It is my kynd to doo soo' and begged to be spared. But the man laughed and said, 'How be it that thow mayst not hurte me ... wherefore thow shalt dye.' The moral is, 'For men ought not to leue none euyll unpunysshed how be but nat greate' (see Fig. 5.8). La Fontayne (1621–95) brings a flea into a moral-pointing tale, which I have very freely translated as:

The man and the flea

> Our importunate clamour doth weary high Heaven
> Often for trifles unworthy of man
>
> A bumptious man felt a prick on his breast
> From a flea hiding deep in his garment's fold.

He cried for attention 'Oh Hercules bold!
You must rid the world of this vicious pest.
And Jupiter, as in the clouds you rest
Will you 'venge me not? Is my cause grown cold?'
To extinguish a flea, he began to scold
The gods, to send lightning at his behest.

Fig. 5.8. Illustration of Aesop's fable of the man and the flea as in Caxton's text.[3]

¶ The xv. fable is of the flee and of the man

He that doth cruyl howe be it that ye euyll be nat great men ought nat to leue hym vnpunyssheD, as it apereth by thys fable of a man which toke a flee. yt bote hym, to whom the man sayde in thys maner. Fle why bitest thou mez latest me nat slepe, & the flee answerd it is my kynde to do so wherfore I pray the yt thou wylt nat putt me to deth & the man bega to laughe & sayde to the fle thou maiste me nat hurt sore. neuertheleffe the behoueth nat to byte me. wherfore thou shalt dye. For men ought nat to leue no euill vnpunysshed howe be it that it be nat greate.

Another versified fable with a philosophical trend came from the pen of John Gay[104] (1685–1732) of *Beggar's Opera* fame. Also entitled *The Man and the Flea*, its theme is vanity in all creatures. Naturally, 'Man, the most conceited creature' surveys the world in his turn and decides:

> ... and all these by Heav'n design'd
> As gifts to pleasure human-kind.

This conceit is too much for a flea, feeding on his nose, who reproves him as follows:

> 'Tis vanity that swells thy mind
> What, heav'n and earth for thee design'd!
> For thee, made only for our need
> That more important Fleas might feed.

This poem might provide a useful answer to those anthropocentric folk who sometimes ask entomologists 'Why were bugs and lice created? What *good* do they do'.

Fleas provided an analogy of a different kind for Jonathan Swift[297] (1667–1745):

> So naturalists observe, a flea
> Has smaller fleas that on him prey;
> And these have smaller fleas to bite 'em,
> And so proceed *ad infinitum.*
> Thus, every poet, in his kind
> Is bit by him that comes behind.

Later in the century, poets took further notice of naturalists, looking askance at scientific curiosity. Pope[250] (1688–1744) for example, in the *Rape of the Lock,* classes

> Cages for gnats and chains to yoke a flea
> Dried butterflies and tomes of casuistry

with suspect items, such as

> The courtier's promises and sick man's prayers
> The harlot's smile and the tears of heirs.

William Cowper[69] (1731–1800) in a long poem on *Charity* felt that science was acceptable if followed in a God-fearing spirit.

> But reason still, unless divinely taught
> Whate'er she learns, learns nothing as she ought.
>
> And without this—whatever he discuss
> Whether the space between the stars and us
> Whether he measure earth, compute the sea
> Weigh sunbeams, carve a fly or split a flea
> The solemn trifler, with his boasted skill
> Toils much, and is a solemn trifler still.

Another taunting poem was directed against Sir Joseph Banks, President of the Royal Society, by Peter Pinder,[242] who was induced to do so (according to Lehane) by jealous members of the Society. The rhyme describes experiments of a mythical Jonas Dryander, to ascertain the suitability of fleas as food, on the grounds of their resemblance to lobsters.

> 'I've just boiled fifteen hundred,' Jonas whin'd
> 'The dev'l a one change colour could I find.'
> Then Jonas curs'd, with many a wicked wish
> And show'd the stubborn fleas upon the dish.
> 'How!' roar'd the President, and backward fell;
> 'There goes, then, my hypothesis to hell!'

Scientific curiosity was abhorred by E. T. W. Hoffman[140] (1776–1822), a versatile genius, but best known to us for his bizarre stories perpetuated in Offenbach's opera. In Hoffman's curious work *Master Flea*, Leeuwenhoek and Swammerdam are derided for infringing the Mystery of Nature.

You poor misguided creatures, your lives have been one continuous uninterrupted illusion. You make bold to investigate Nature, but had no conception of her innermost meaning ... by embarking on a wanton enquiry into every detail of the causes and condition of those wonders, you destroyed that reverance and the knowledge after which you strove.

Master Flea of the story is the leader of a lilliputian nation struggling for freedom, who seeks aid from one Peregrinus Tyss, who is the real hero. One of the episodes (not involving the flea) satirizes the 'Demagogue trials' of the time in which democratic leaders were persecuted by the reactionary authorities.

By the nineteenth century, references to insect vermin in poetry tend to diminish, but sometimes they turn up in fables and jokes. Well known, of course, is the 'Song of the Flea' from *Faust* by Goethe[110] (1749–1822).

> A king there was, be't noted
> Who had a lusty flea
> And on this flea he doted
> And loved him tenderly.
> A message to the tailor goes
> Swift came the man of stitches
> Ho, measure this youngster here for clothes
> And measure him for breeches.

> In silks and satins of the best
> Soon was the flea arrayed there
>   Ribbons he had his breast
> Likewise a star displayed there.
> Prime minister he grew anon
> With star of huge dimensions
> Got title, rank and pensions.
> And lords and ladies, high and fair
> Were grievously tormented
> Sore bitten the queen and her maidens were
> But they did not dare resent it.
> They were afraid to scratch
> Howe'er our friend might sack them
> But we without a scruple catch
> And when we catch, we crack them.

Mephistopheles sings this song to beguile a cellar full of merry drinkers, presumably as a jest. But the analogy might serve well for the rise of many a king's favourite.

Fleas appear in two stories of Hans Andersen[11] (1805–75). *The Flea and the Professor* concerns the highly fantastic adventures of a performing flea and his trainer. The other is the tale of the *Three Leapers;* a flea, a grasshopper and a skipjack (which is a kind of jumping toy). In a competition to win the hand of a king's daughter, the flea leaps highest, but being too small, no one can see how high. The skipjack wins by leaping on to the Princess's lap, the highest place to which anyone could aspire.

Later in the century appeared a curious little poem in a book devoted to the droller, not to say Rabelaisian aspects of medicine: *Parnasse Hypocratique*[218] (1884). (It contained verses entitled 'Etes-vous circoncis?', 'Desir de femme grosse', 'La confession d'une nymphomane,' etc.) The one which concerns us concerned Pope Sextus V (1521–90), who rose from humble origins, and is the only poem I know celebrating *Pthirus pubis,* which I have freely translated:

> Pope Sextus wearing on his brow
> Of healer, priest and king, the crown
> Gazed from his high window down
> Enjoying the peaceful evening. Now
> Was heard the whisper of a pilgrim louse
> From some deep private lair
> Mounting familiarly to say
> Remember'st thou that far off time

When we, poor wretches, tended swine,
We whom the world reveres today?
Ah! worthy he of Herculean fame
Who could detach me from thy noble frame.

Attempting to divine a possible hidden meaning in this verse, I conclude that it could perhaps be interpreted as an analogy to the close attachment of human parasites to those who rise to eminence.

Insect Vermin and Weather Vanes

A colleague returning from a tour of north Hampshire, told me of a curious, persistent legend in the little ancient town of Kingsclere, to the effect that the vane on the old church represented a bed bug. According to local tradition, King John (who certainly visited the neighbourhood on many occasions) once spent an uncomfortable night with the local monks. In ironic gratitude, he presented them with a weather vane in the form of a large iron bug (Fig. 5.9*a*). If this story were true, it would point to a very early introduction of bed bugs in England, for John died in 1216. Unfortunately, further enquiries revealed several flaws in the legend. According to the city librarian at Winchester, the church tower at Kingsclere was raised during the fourteenth century and it was fitted with a new vane in 1848. The one which it replaced is said to have represented a tortoise, but this seems unlikely, particularly as there are six legs. In 1905, the Rev. Finch,[90] who was then vicar, published an account of 'The old Church at Clere'. He does not mention the bed bug story; but perhaps this was from delicacy.

While I was considering this story, I read of another old weather vane in London.[12] The church of St Luke, Old Street, has apparently long been known as 'lousy St Lukes' on account of a weather vane said to represent a louse (Fig. 5.9*b*). According to a local story, when the church was built in 1733, the builder—disgusted with the parsimonious parish officials—put up the louse weather vane as a sign of contempt. The vane was blown down by bomb blast in 1940 and when the vicar decided to restore it, he consulted an expert from the Guidhall Museum, who pronounced that it represented a dragon. If, indeed, dragons are to be considered as possible motifs for weather vanes, I believe the one at Kingsclere could be a crude copy of one of Aldrovandi's[7] monsters Fig. 5.9*c*).

Mr Norman Clark of the Guildhall Museum has also informed me of a third example of this curious association of vanes and vermin. The church of St Olave and St John in Southwark, had a vane which was known locally as the 'louse'. It was, in fact, the Sun in its Glory, to use the heraldic term. This church also suffered from air raids, but I have copied the vane from an old print (Fig. 5.9*d*).

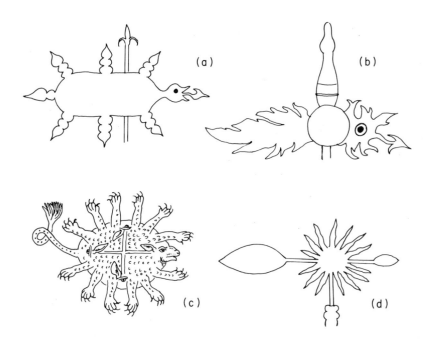

Fig. 5.9. Vermin and Weather Vanes. *a*, the weather vane at Kingsclere; *b*, the weather vane at St Luke's, Old Street, London; *c*, Aldrovandi's monster;[7] *d*, the weather vane at St Olave and St John, Southwark.

These three cases of identifying church vanes with insect vermin seem to require some explanation. Possibly there is some curious, forgotten tradition to account for them; but perhaps there is a more prosaic reason. A vane on a steeple seen from the ground is a small black object. If it should be decorated with anything vaguely resembling legs it is not unlikely that an association with lice or bugs would occur to the more ribald members of the congregation.

Lice and Love: Fleas and Fleshy Fantasies

A medieval riddle, attributed to Konrad von Megenberg (1309–74) runs:

What animal is most faithful to man? The louse, which once attached to a man stays with him till death.

On the same lines, a character in the *Merry Wives of Windsor* says of the louse:

It is a familiar beast to man and signifies love.

We find the same joke still flourishing three centuries later, when a contributor[1] to the Gentleman's Magazine in 1746 made oblique references to the louse as a creature noted for its affection for man, its fidelity, its modest and retiring nature and so forth.

Somewhat more salacious fancies concerning lice and fleas derive from their freedom to crawl over the most intimate corners of the human frame. It is therefore not very surprising to find Rabelais[257] (1483–1553) describing some slightly sexy practical jokes enjoyed by Panurge.

In another . . . he had a great many little horns full of fleas and lice, which he had borrowed from the beggars of St Innocent, and cast them with small canes or quills to write with, into the necks of the daintiest gentle-women that he could find, yea even in the church.

And again:

. . . a fair handkerchief, curiously wrought, which he had stolen from a pretty seamstress of the palace, in taking away a louse from off her bosom, which he had put there himself.

The sexy adventures of the pedestrian louse, however, are nothing to the frisky flea. The Clown in Marlowe's[198] *Faust* begs Wagner:

If you turn me into anything, let it be in the likeness of a pretty frisking flea, that I may be here, there and everywhere, O I'll tickle the pretie wenches' plackets I'll be amongst them i'faith.

The impudent flea is referred to again in a later scene, where Pride remarks:

. . . I am like to Ouid's Flea, I can creepe into euery corner of a Wench: sometimes like a Perriwig, I sit upon her Brow: next, like a Necklace, I hang about her Necke, Then, like a Fan of Feathers I kisse her: And then, turning muselfe to a wroughte Smocke, I do what I list.

Marlowe is referring to a poem written in the style of Ovid, probably in the late Middle Ages. This ribald poem was popular among the erudite of the sixteenth century; even sober Mouffet makes passing reference to 'the wanton poet' in discussing fleas. In an essay on 'Ovid in the Middle Ages' Dorothy Robathan[268] remarked: 'In view of Ovid's popularity in such diverse fields of learning, it is perhaps not surprising that he was often imitated by mediaeval admirers, some of whom had no scruples in putting his name to their own compositions'. F. W. Lenz[183] has made a very careful study of the origins of '*De Pulex libellus*', which he has traced back to the twelfth century. He notes that one of the popularizers of the spoem, the Swiss historian and publicist Melchior Goldast (1576–1635) ascribed it to '*Ofilius Sergianus*'; but no such person is known and the name may have been invented, (possibly with *Ofilius* to rhyme with *Ovidius*). It runs:

> *Parve pulex sed amara lues, inimica puellis*
> *Carmine quo fungar in tua facta ferox* . . . etc.

which I have freely translated, as follows:

Oh, tiny flea, the maiden's foe
Who shall sing of your deeds of woe?
A tender body you ill-treat
Which, when you leave, with blood replete
Is mottled o'er with dusky stains
Unlucky limbs, the trace remains.
And when you plunge you hidden beak
You wake the troubled maid from sleep.
Amid the folds, all limbs to view
You wander free; nought hid from you.
Ah shame! A girl asleep is seen
You part her legs and creep between.
Sometimes you even dare to vex
The parts that hide the joys of sex.
What gall! My enemy can go
Along the road I yearn to know.
Yet I would do what you can not.
Could I regain my human lot
After a change, my choice would be
To metamorphose to a flea.
By magic, or if drugs could do it
I'd change my shape and never rue it.
Medea's spell or Circe's lotion
Conjoined into a magic potion.
Thus changed, if such a change were known

I'd creep upon the maiden's gown
Then quick beneath her shift I'd sink
And to that place—right to the brink
Then, harming no one in that bed
I'm flea no more, a man instead.
But, if the girl took fright at last
And called her maids to bind me fast
Either by prayers she'd yield to me
Or else I'd change back to a flea.
Changed yet again, with many a vow
The Gods I'd summon to help me now
Until by force or prayer, my end
I gain, and I'm the girl's best friend.

Lenz points out that the poem is almost certainly inspired by a genuine Ovid elegy to a Ring (*Amores* II.15). A suitor sends his lady love a ring and wishes that he could transform himself into his gift. Close embracing her finger, he would wish her to lay her hand beneath her tunic and he would contrive to slip off into her bosom. Then, if she wore the ring in her bath, the vision of her beauty would drive him to play the human part *(Sed puto, te nuda mea membra libidine surgent)*.

It is amusing to contrast the ways in which this theme was treated in France, England and Germany. In Poitiers in the latter half of the sixteenth century, two beautiful and gifted women held court to some of the more dashing intellectuals of the time. They were Madelene Neveu, widow of the seigneur Des Roches and her 'fille d'alliance' Catherine des Roches. Among the members of their salon were Etienne Pasquier, poet, writer and member of Parliament, Claude Binet, friend of Ronsard, lawyer and poet and Nicholas Rapin, also a poet and, incidentally, Grand Provost of the Constabulary.

One day in 1788, at one of their gatherings, one of these gallants noticed a flea on the bosom of the charming Catherine; some innuendoes followed, no doubt, and finally a whole series of little poems were composed on the subject. The general tone can be judged from the opening lines of some of these verses.[233]

La Puce de Catherine des Roches	La Puce d'Etienne Pasquier
Petite Puce fretillard	Puce qui te viens percher
Qui d'une bouchett mignard	Dessus cette tendre chair
Sucotes le Sang incarnat	Au milieu de deux mammeles
Qui colore un sein delicat	De la plus belle des belles
Vous pourroit-on dire friande	Qui la piques, qui la poignts
Pour desirer telle viande?	Qui la mors a tes bon poincts

La Puce de Claude Binet

Mignard, vous avez grand tort
D'appeller Hercule a la mort
A la mort d'une pucelette
Qui tout mignardement furette
Comme un petit surjon d'Essain
Sur les roses vostres sein

Contre-Puce de Nicholas Rapin

Puce que tant de bon espris
Pour sujet de leurs vers en pris
Qui t'ont trouvee, si habille
Que, la Muse les enchaufant
Ils t'ont fair un grand Elephant
Par leur invention gentille

It seems highly probable that these verses prompted the anonymous seventeenth century poet quoted by Lehane:[182]

> Madam that flea that crept between your breasts
> I envied, that he there should make his rest:
> The little Creature's fortune was so good
> That Angells feed not on such pretious foode
> ... etc.

Possibly, too, they may have inspired John Donne[79] (1573–1631) for a slight variation on the theme:

> Marke but this flea, and marke in this
> How little which thou deny'st me is,
> It sucked me first, and now sucks thee.
> And in this flea our two bloods mingled be.
> Thou knows't that this cannot be said
> A sinne, nor shame, nor losse of maidenhead,
> Yet this enjoys before it wooe
> And pampered swells with one blood make of two
> And this, alas, is more than wee would doe.

The German satirical poet, Johann Fischart[81] (1545–1614), whom I have mentioned earlier, dealt with other aspects of feminine flea infestation, than the purely erotic. His poem *Flöh Hatz/Weiber Kratz* published in 1573 concerns a verbal duel between women and fleas, as to the right of the latter to take human blood. Jupiter finally gives judgment that fleas may only drink human blood at dire peril of being caught and killed. However, he allows them three opportunities of feeding on women: on their tongues (with which they belabour their menfolk), round their puffered out ruffs at hand and neck, and at the dance. The writer seems to have been slightly misogynistic. The title page (reproduced in Fig. 4.2) shows the effects of the fleas on various exasperated women. A variant of this story comes into *Simplicissimus* by Grimmelshausen[13] (1625–76). This is a rambling account of adventures in some ways reminiscent of Pantagruel and Till Eulenspiegel; but it gives some convincing pictures of life during the 30 years' War. In Book III Ch. 6, we meet Jupiter, greatly troubled by fleas, who had come to him to complain of the way they were being persecuted by women. They 'complained to him that, though he had assigned to them dog's coats as a dwelling, yet on account of certain properties common to women, some poor souls went astray and trespassed on the ladies' furs; and such poor creatures were by the women evil entreated, caught and not only murdered, but first so miserably martyred and crushed between their fingers that it might melt a heart of stone.' When Jupiter pointed out that the women, as hunters, were entitled to snare their game, the fleas begged at least to be 'executed in honorable wise, and either cut down with a pole-axe, like oxen or snared like game . . .' So Jupiter takes pity on them and allowed them to lodge with him; but later regretted it!

In the eighteenth century, that wordly dilettante, the abbé Willart de Grecourt,[322] (1683–1743), was still exploiting further variations of the intruding intimacies of 'Ovid's flea'. In one poem, a flea boasts to a rat that

> . . . *tres souvent, sans crainte*
> *Jusque dans le pays d'Amour*
> *Elle parcourut toute Aminte.*

('Elle' being the flea and 'Aminte' the human landscape). Another flea, boasting to a sparrow this time, that

> . . . *les endroits qui convoitent les Dieux*
> *Ne sont pour moi sacres ni precieux.*

The fascination of the erotic peregrinations of fleas was still present in the nineteenth century, as witness a poem by Wilhelm Busch[45] (1832–1908). Born in Hanover, he studied painting in Düsseldorf but was mainly renowned for his cartoons, to which he added droll verses.

Der Floh

Der Abend is so mild und schön.
Was hört man dafür ein Getön?
Sei ruhig, Liebchen, das bin ich,
Dein Dieterich.
Dein Dietrich singt so inniglich.
Nun kramst Du wohl bei
    Lampenschein
Herum in Deinem Kämmerlein;

The Flea

The evening is so mild and clear
What trifling sound assails the ear?
Don't worry, Darling, it is I,
Your Derrick
Your Derrick sings so soulfully.

The lamp dispels the evening gloom
You rummage round your little room

Nun legst Du ab der Locken Fülle,	Now remove your ample hair
Das Oberkleid, die Unterhülle;	Your outer clothes, your underwear
Nun kleidest Du die Glieder wider,	Then clothe your limbs in sleeping gown
In reines Weiss und legst Dich nieder.	Of purest white, and lay you down
Oh, wenn Dein Busen sanft sich hebt,	Oh, as your bosom swells, my dear
So denk, dass Dich mein Geist umschwebt:	Think of my spirit floating near.
Und kommt vielleicht ein kleiner Floh	Perchance there comes a little flea
Und krabbelt so—	And gently crawls—
Sei ruhig, Liebchen, das bin ich	Don't worry, Darling, it is I
Dein Dietrich	Your Dietrich.
Dein Dietrich der umflattert Dich	Your Dietrich hovers nigh.

An ironic chapter by Lehane[182] describes the further exploits of the erotic flea in the eighteenth and nineteenth centuries, which were somewhat repetitive and lacking in taste. 'The twentieth century', he notes, 'needs wrier stimulants' and concludes, 'Flea pornography is a microcosm of pornography in general. It offers a deal of indecent exposure without any striking revelation. For centuries, however, it occupied the fabled flea, made it aware of human traits and frailty. It is a chapter in the flea biography, and the chapter is done.'

Epitaphs for Insect Vermin

A curious medieval pseudo-Ovidian poem of the early thirteenth century records the fate of a louse. It has been closely connected in manuscripts with the somewhat salacious poem about a flea (p. 114) and is likewise the subject of a lucubration by Dr F. W. Lenz.[184] It begins:

> *In cute sudanti sub veste pediculus hesit*
> *atque cutem rupit,suxit et intumuit . . .*
> Beneath my clothes, on sweaty skin
> A sucking louse began to swell
> I felt it prick—it made me itch—
> My eager finger sought it well.
> Brought to the light, it fled my thumb
> The wriggling slipping creature fell
> Between my knuckles; but in vain
> Soon sought and caught
>   to trial, a prisoner brought
> Condemned to suffer death in pain.

His sisters and their nits bewail his doom
And make this epitaph upon his tomb.
A HEAD WITH SIX FEET AND BODY WITH NO BREAST
LIES IN THIS TOMB, FLESH MADE FLESH FROM FLESH.

My other two examples come from a monograph by G. Herrlinger,[129] *Totenklage um Tiere in der antiken Dichtung*. In addition to Egyptian, Greek, Latin and Byzantine epitaphs, he includes some German ones up to the nineteenth century, for comparisons of style. Herrlinger divides them into genuine expressions of sentiment (as, for example, laments for favourite dogs), parodies and epigrams. The two I propose to quote concern fleas, written in a humorous-satirical style, somewhat on the lines of Martial's epigrams. The first is by Chr. Hofmann von Hofmannswaldau (1617–79)

Ein schwartzer Rittersmann fiel durch ein weisses Weib
In dem er ohne Scheu betrat den zarten Leib
Doch is sein alter Ruhm nicht gantz und gar verdorben
In dem er eben so wie Curtius gestorben
A black knight slain by a fair white maid
Whose tender skin he dared invade
Nor is the fame of his deed grown cold
Who died like Curtius of old.

Curtius was a Roman knight who leapt into a chasm in the forum, as a kind of sacrifice.

The other example, by 'Celander' (apparently Christopher Wottereck, 1686–1735) is even more indelicate.

Nachdem ich lange Zeit die weisse Brust bewacht
So kahm ich an den Ort wo selbst die Anmuth lacht,
Allein den Vorwitz must ich mit dem Leben büssen
Denn allda wil man nichts als grosse Stachel wissen.
On a snow-white breast I tarried a while
Then came to the spot where pleasures smile,
For rashness now my life is ceded
Since here much bigger pricks are needed.

Fleas as Puppets and Performers

For various reasons, we find some wild animals much more attractive than others. Even among the pests there are favourites. Mice and rats are both annoying rodents; but whereas rats are generally detested, mice have been the heroes of divers comic strips and cartoon films (remember Teddy Tail, Mickey Mouse, Toppo Gigio, Jerry?). Even at the level of ectoparasites, there are quite different reactions; thus, my quotations suggest that our ancestors regarded fleas with tolerant amusement, whereas lice and bugs merely aroused aversion. Kirby and Spence[169] (1815) quote the remark of a lively old lady to a friend who was confined to bed with a broken limb and complained of fleas. 'Dear Miss, don't you like fleas? Well, I think they are the prettiest little merry things in the world . . . I never saw a dull flea in all my life!' Even the eminent German zoologist, A. E. Brehm,[34] (1829–84) includes a mildly approving rhyme in the flea entry of his 10-volume *Tierleben*.

> *Glücklich drum preis' ich den lockeren Gesellen*
> *Pulex, den Turner im braunen Trikot*
> *Wenn er in Sprüngen, vergangen, schnellen*
> *Himmelhoch jauchzet, frisch, fromm, frei und froh!*
> That raffish fellow I gladly praise
> Pulex, the gymnast in sleck brown tights
> Spontaneous springs will readily raise
> His joyful jolly jumps to heavenly heights!

The minute size of fleas offers a challenge to intricate workmanship and, furthermore, there is an anthropomorphic desire to create a human world in miniature. According to Lehane,[182] Mexican nuns used to dress dead fleas in tiny costumes and these *pulgas vestidas* were sold as souvenirs. There are two examples of this curious art in the flea collection at Tring started by Charles Rothschild before the first World War; and an even more elaborate exhibition is on show at the Museum of Childhood in Edinburgh.

If dead fleas are interesting, live ones can be still more fascinating and there are even records of them being kept as quaint pets. It seems that French gallants of long ago used to cherish a flea caught from their lady love. It would be kept in a kind of reliquary on a gold chain hung round the neck, and fed daily. The late Professor de Feytaud[89] stated that he had read that the poet and libertine Jacques Vallee Des Barreaux (1602–73) had thus cherished a flea caught on Marion Delorme (1612–50) a celebrated courtesan. Mouffet[219] mentions 'one *Mark* an *Englishman* (most skilful in all curious work) fastened a Chain of Gold as long as a man's finger, with a lock and key so rarely and cunningly, that the Flea could easily go and draw them, Yet the Flea, the Chain, lock and key were not above a grain weight: I have also heard from men of credit, that this flea, so tied with a Chain, did draw a Coach of Gold that was in every way perfect, and that very lightly; which much sets forth the Artists skill and the Fleas strength.'

A century later, John Ray[261] the naturalist was travelling in Europe with his friend Francis Willoughby, through Venice and Augsburg, when they met 'one who sold fleas with steel or silver chains, one of which Dr Willoughby obtained. This was kept in a box in a warm place; and allowed to feed daily, it lived a long time'. In *Histoire abregee des Insectes qui se trouve aux environs de Paris* by E. L. Geoffroy[106] (1725–1810) we read '*Hoock [sic] raconte un fait encore plus surprenant. Un ouvrier anglais avoit construit en ivoire un carosse a six chevaux, un cocher sur la siège avec un chien entre ses jambes, un postillon, quatre personnes dans la carosse et deux laquais derrière, et tout cet équipage étoit trainé par une puce.*' I cannot, however, trace this in any of Robert Hooke's writings; Geoffroy cites the *Micrographia*,[143] but it certainly is not there, nor in *Lectures and Collections*.[144]

In Hoffman's[140] story of Master Flea (which I have mentioned earlier, p. 109) there is an account of a 'fleamaster' (*Flöhbander*) who drills an army of fleas on a table top. They carried minute muskets, cartridge cases and sabres and their manoeuvres were open to the public and attracted many viewers. About this time public flea shows were fact as well as fiction. Kotzebue[173] (1761–1819) an extremely prolific German dramatist recorded one in his *Erinnerungen aus Paris* in 1804. It seems that a sailor was exhibiting under a magnifying glass a flea pulling an elephant appropriate to its size and another drawing a tiny carriage. From this time onward, there are a series of records. Baron de Walckenaer[314] (1771–1852) who lived precariously through the French Revolution and later devoted himself to literary and scientific pursuits, refers to a flea exhibition in his *Histoire Naturelle des Apteres* (1844). Once again we have the fleas harnessed to a carriage and two other pulling a tiny cannon. Thirty others engaged in fencing with minute splinters. The fleas, he notes, were fed on a man's arm and when they appeared sluggish in their performance, were warmed up by waving a hot coal near them. Reference to such performances are made by Brehm[34] in *Tierleben* and even more detailed descriptions in the French edition in the series by Kunckel d'Herculais. Somewhat later in the nineteenth century Gaston Tissandier,[301] who wrote popular science articles on ballooning and so forth describes a similar exhibition in *Recreations Scientifiques* (1876). Finally, Theodor Birt, an antiquarian and scholar wrote of them in his memoirs of childhood. It seems that, at the annual *Domfreuden* or fair in Hamburg, among the jugglers, fire-eaters and 'fattest men in the world' would be found a Flea Circus. This tradition has continued until modern times and Prof. Weidner[317] publish-ed photographs of the 1965 version. Alas, however! The beautiful metal workmanship of the old flea exhibitions has degenerated to paper and cardboard; and the panache and showmanship of the early inhibitors has come down to a tape-recorded loudspeaker! Memories of these honorary Professors survive in anecdotes and in some of the adver-tisements for their shows.[89] About 1834, there was Cucciani who dressed his performers as soldiers and gave them commands, while others 'danced' to music. There was Obicini, also in Paris, with his in-evitable joke. 'This flea, ladies and gentlemen, has a famous history. His great, great grandmother was performing in front of the queen and her ladies, and escaped. The ladies searched diligently and finally the queen produced an insect in her royal fingers. I examined it and, bowing respectfully, declared: Alas! Madam, it's not the one.' In London, in the 1830s, there was an outstanding showman in the person of the Italian Bertolotto, who exhibited in the Cosmorama Rooms in Regent Street. He advertised his performance as 'under the patronage of her Royal Highness Princess Augusta' (daughter of George III). Lehane[182] quotes from the programme notes of a 'ball at which flea ladies partner their frock-coated gentlemen, and a twelve-piece flea orchestra plays audible flea music, while in an alcove four whiskery old flea bachelors make up a four at whist. Another scene, a mail coach and coachman in the royal flea livery, belabouring his four flea chestnuts with an actually cracking whip. Fanfares, and a Man of War is drawn on: a hundred and twenty guns, and the whole lifelike miniature drawn by a single flea, though the tableau is four hundred times its weight. Then the Great Mogul, com-plete with harem, a splendid palanquin, and a hookah at which he realistically puffs. Finallly, the grand climax. A hush; and then enter ac-curate portrayals of the three heroes of Waterloo—Wellington, Napoleon and Blücher—three immortal miniature warriors.' Then there was 'Professor' Leidersdorf of Hamburg, whose entire company deserted him in 1853. Overcome with grief, he committed suicide to the great sympathy of the inhabitants of that city.

Towards the end of the nineteenth century, an impressive exhibition was presented in Paris and other European cities by Charles de Wagner. Feytaud[89] reproduces two of the posters advertising the show, from which we read of 300 flea artists. These fleas pulled carriages, wagons, cannons and drive them themselves. There were flea tram conductors, bicyclists, jugglers, tightrope walkers, duellists and so forth. The public were guaranteed against desertions! Feytaud mentions a similar show in Paris in 1923, run by a Mme Stenegry. On being asked how they were fed, the lady called out 'Marie!' and a dejected looking woman appeared and resignedly rolled up her sleeves.

Although Weidner cites the Hamburg flea circus of 1966 and Lehane describes one run by a Professor Tomlin in Manchester's Belle Vue Park, it is evident this is a moribund form of entertainment. To some extent, this is due to more sophisticated (and, perhaps, more squeamish) modern taste; but there is now a real difficulty in obtaining actors. At the end of the nineteenth century, de Wagner offered one franc a dozen for new recruits (human fleas only); but in recent years Tomlin's offer of a pound a dozen failed to obtain supplies. I myself can remember the visit of a flea circus proprietor to our Department offering a good price for 'good bold fleas'. Unfortunately, we could only offer the puny plague flea, *Xenopsylla cheopis* which, like the Red King was 'good enough, but not strong enough'. As a result, flea circuses are disappearing. The one at Hubert's Museum, near Times Square in New York, closed about 1955; another famous one in the Tivoli Gardens, Copenhagen, survived only ten years later.

In conclusion, it is hardly necessary to add that 'performing' fleas are not trained; they are merely forced to pull carriages or wave their swords by appropriate tying or glueing. This ill treatment was actually cited by a deputy to the French Chamber in 1923, during a debate on cruelty to performing animals. But what could express our sympathy better than Ruth Pitter's[244] *Sad Lament of a Performing Flea*?

> Lord! the hard service of the fickle world!
> Ah heaven! to hear of such a thing as ease!
> Envy me not, 'tis I who envy you;
> Beauty avaunt, thou liest and dost betray
> Were I to be new-hatched, ah, then I'd choose
> A poor estate, a leaden louse I'd be;
> They dance not, and remote from public shows
> Pursue their innocent and homely joys;
> ... with a perfect hatred I contem
> My vaunted shape, my thin and polished sides,
> My lovely slender legs that worked my bane
> And these great eyes that see the world too well.

My dwelling fabulous of crystal glass
To me the temple of foul luxury
And weary pleasure is, where I must do
Nothing but dance until I be foredone.
All happy things that pasture where you please,
For my much-loathed diet pity me,
That only feed upon the hairy arms
Of my cruel tyrants keeping me for gain;
I that was born to range the merry world
To rob at will the veins delectable
Of princes, and taste the wholesome blood
Of sweet-breathed peasants, skipping light aside
With silver laugh at scratch perfunctory;
To lie with ladies, and ah fairest joy,
On infants' necks to feed, until a bee
I seem that rifles a white breathing rose.
Example take by me; and O complain
Never too loudly of a low estate . . .

6 Ectoparasites as Objects of Scientific Curiosity

Classical Times

The flowering of philosophical and scientific speculation among the Ionian Greeks was almost unique. They were interested in things for their own sake, rather than for practical reasons, and this even applies to such lowly things as insect vermin. A passing mention of ways of eradicating such things has been noted in early Egyptian writings (p. 67) as well as in later Roman records, while early physicians have suggested their use as medicines (p. 176). The ancient Jews were concerned with their religious and legal aspects. The first evidence of a purely scientific interest in animal life comes not meagrely but in full flood, with Aristotle (384–322 B.C.). He was, I suppose, the archetypal academic, beginning as a pupil of Plato and becoming tutor to the young Alexander of Macedon. When the latter started his career of conquest, Aristotle moved to Athens; here he became a public teacher, in his garden the 'Lyceum' where he established his famous 'peripatetic' school, from his habit of walking around while lecturing. Although he had something to say on all aspects of knowledge, it is his biological writings which are outstanding. The vividness and general accuracy of many of his statements suggest that they were made from personal observation. Yet his material is so prolific that he must often have relied on reports of others; for example, he records the life history and breeding habits of about 540 species of animals, including such exotic beasts as the elephant. Bertrand Russell[272] has remarked that there are two ways of regarding this unique man. As a genius, he towers above his contemporaries, for there were few philosophers and virtually no biologists to equal him for centuries. On the other hand, his very eminence, combined with the narrow religious and authoritarian outlook of the middle ages, stifled all criticism and further thought, so that even his more easily demonstrated errors went unchallenged for over eighteen hundred years. These two aspects of Aristotle are exemplified by his writings on the insect parasites of man. In his *Historia*

Animalium,[15] he gives the following account of them (v.30): 'Of insects that are not carnivorous, but live on the juices of living flesh, such as lice and fleas and bugs, all without exception generate nits, and these nits generate nothing. Of these insects, the flea is generated out of the slightest amount of putrefying matter; for wherever there is any dry excrement, a flea is sure to be found. Bugs are generated from the moisture of living animals, as it dries up outside their bodies. Lice are generated out of the flesh of animals. When the lice are coming, there is a kind of small eruption visible; and if you prick an animal when in this condition at the site of eruption, the lice jump out. In some men, the appearance of lice is a disease, in cases where the body is surcharged with moisture; and, indeed, men have been known to succumb to this louse-disease, as Alcman, the poet and the Syrian Pherecydes are said to have done. Moreover, in certain diseases, lice appear in great abundance. There is also a species of louse called the wild louse and this is harder than the ordinary louse, and there is exceptional difficulty in getting rid of it. Boys' heads are apt to be lousy, but men's in less degree; and women are more subject to lice than men. But whenever people are troubled with lousy heads, they are less liable to headaches. And lice are generated in other animals than man, including birds . . . and all hair-coated creatures also, with the single exception of the ass, which is infested neither with lice nor ticks.'

Aristotle's genius is evident from the very fact that he bothers to record observations and make speculations about creatures that other ancient writers considered beneath contempt, except as a term of abuse or subject for a vulgar joke. Some of his statements were erroneous, but they are not foolish. To make a comparison: when Dioscorides[77] (*c.* 50) claimed that bed bugs taken medicinally would cure quartan malaria, he was talking nonsense; but when Aristotle stated that fleas are generated from excrement, he was merely guilty of insufficiently thorough observation. Furthermore, Aristotle's biology was scarcely improved upon for about 2000 years; so that, in the sixteenth century, his words were still being quoted extensively as major portions of zoological treatises. At the time of the Renaissance, his errors and ambiguities were causing considerable trouble (as we shall see) when scholars tried to reconcile more or less distorted versions of his texts with what they actually saw. The matter can be well illustrated by various passages on the subject of ectoparasites. Thus, the statement about pricking small pustules and releasing 'lice' has given trouble and some authorities believe that scabies mites were meant. The practice of pricking out mites seems to be well known among peasants in several lands. On the other hand, it is difficult to be sure. 'Wild' lice also have caused confusion, though the most likely explanation is that he was referring to crab lice.

Aristotle's remarks about the prevalence of lice in young people as opposed to old and on women rather than men are perfectly sound (at least, in regard to head lice). Unforunately, in *Problemata*[15] (II.16) he indulges in the very human pleasure of trying to explain it. 'The brain is moist and therefore the head is always the moistest part of the body as shown by the fact that hair grows there more than elsewhere, and it is the moisture of this part that generates lice. This is clear in the case of children; for their heads are moist and they frequently have either running noses or discharge of blood, and persons of this age suffer particularly from lice.'

To return to the *Historia Animalium*, the erroneous statement that asses are never infected with lice was accepted without question for centuries. It agreed well with a religious fable of divine dispensation on the grounds that Christ rode on an ass into Jerusalem. Aristotle's error was not challenged until, in 1668, Francesco Redi published a considerable number of figures of biting and sucking lice from different animals, including one from the ass. But as late as the end of the sixteenth century, Thomas Mouffet was supporting Aristotle with the suggestion that, since asses are notoriously lazy, they seldom sweat and hence do not provide matter for the generation of lice. Then, possibly thinking this explanation specious, he adds the medieval argument of antipathy—the last refuge of logic, as Redi commented. A more important restrictive influence of Aristotle was the support which he gave to the doctrine of spontaneous generation. In *De Generatione Animalium,* Aristotle deals with insects as follows (I.16). 'Some insects copulate and the offspring are produced from animals of the same kind, just as in the sanguines; such are the locusts, cicadae, spiders, wasps and ants. Others unite and generate; but the result is not a creature of the same kind, but only a *scolex,* and these insects do not come into being from animals but from putrefying matter, liquid or solid; such are fleas, flies and cantharides. Others again are neither produced from animals nor unite with one another; such are gnats and many similar kinds. The female is larger than the male. The males do not appear to have spermatic passages. In most cases the male does not insert any part into the female, but the female from below upwards into the male; this has been observed in many cases (as also that the male mounts the female) the opposite in a few cases . . .' (Notes: *Sanguinea* or animals with blood are the larger animals—vertebrates. A scolex is apparently either an egg or a pupa.)

The next important name is that of Pliny the Elder (23–79). Born at Como, he spent his youth in Rome, where he became interested in plants, though later turned to philosophy and rhetoric. Military service took him to Germany and also to France and Spain. On his return to Rome, he wrote his enormous *Natural History*,[246] a compendium drawn from about 2000 works, by 326 Greek and 146 Roman authors. On friendly terms with Vespasian and Titus (to whom he dedicated the *Natural History*) he was appointed an admiral of the West Mediterranean fleet. In this capacity he lost his life at the eruption of Vesuvius which overwhelmed Pompeii and Herculaneum, which he was endeavouring to observe too closely.

Pliny's[246] remarks on the parasites of man are as follows (II.39): 'Then too in dead carrion there are animals produced and in the hair of living men. It was through such vermin as this that the Dictator Sylla and Alema, one of the most famous of the Greek poets, met their deaths. These insects infest birds, too, and are apt to kill the pheasant, unless it takes care to bath itself in the dust. Of the animals that are covered with hair, it is supposed that only the ass and the sheep are exempt from these vermin. They are produced, also, in certain kinds of cloth, and more particularly in those made of the wool of a sheep which has been killed by a wolf. I find it stated by authors that some kinds of water which we use for bathing are more productive of these parasites than others. Even wax is found to produce mites which are supposed to be the very smallest of living creatures. Other insects, again, are produced by the rays of the sun—these (fleas) are called *petuaristae* (leapers) from the activity they display with their hind legs. Others again are produced with wings, from the moist dust that is found lying in holes and corners.' This section is not especially impressive, bearing evidence of dependence on Aristotle and of incorporating unlikely folklore. Pliny has, indeed, been castigated as a mere uncritical compilator, though perhaps this is too severe a verdict. The particular passage I have quoted is not among his best. The introductory sections in Book II concerning the insects, seem to show signs of critical judgment and even of personal observation. Thus, he notes that several writers have denied that insects have blood or respiration; but he feels that creatures, many of which inhabit the very element of respiration, must employ it. As for blood, he thinks they have something similar, by way of equivalent. Then '. . . except in some few, which have an intestine, arranged in folds. Hence it is that, when they are cut asunder, they are remarkable for their tenacity of life, and the palpitations are to be seen in each of their parts.' This somehow seems to me to indicate, personal observation, not to be despised in one without even a hand lens. He grants them, too '. . . eyes and the senses of touch and taste; some of them also have the sense of smelling and some few that of hearing.'

Whatever one's assessment of Pliny, he was certainly the most outstanding Roman authority on natural history. As Charles Singer[288] points out, 'all educated Romans came to learn Greek and were inevitably affected by Hellenic philosophy. Yet despite the stimulus of Alexandrian thought, the Latins produced no great creative men of science.' This, he suggests, is a consequence of their adherence of the stoic creed, which led to '. . . either a type of exact but intellectually motiveless observation, or a rejection of all knowledge not of practical importance'.

Pliny himself was considerably interested in the practical importance of insects; and this is even more the case with other Roman writers, who concern themselves extensively with honey bees and a little with agricultural pests, but scarcely with their natural history. Other writers of the time deal with quasi medical aspects of ectoparasites. The Roman physician Celsus (first century A.D.) describes scabies (though not the itch mite) and gives an ingenious method of extracting a flea from a patient's ear! Likewise, in the eastern Mediterranean, Dioscorides the herbalist (first century A.D.) and Galen the physician both mention lice and bugs, but do not contribute to their biology.

The Dark Ages

If inspiration and innovation failed in the Roman Empire, even worse was to follow. When the Western Empire collapsed under barbarian invasion, virtually all intellectual scientific activity ceased. In the succeeding Dark Ages (say, from 400 to 1000) the pervading influence of the Early Christian Church was entirely hostile to scientific curiosity and indifferent to the preservation of the works of heathens such as Aristotle, who seems to have been largely forgotten, and Pliny who, however, fared a little better. Some modern scholars have found it difficult to believe that their works were so neglected; but that is a conclusion one must draw from the rare encyclopaedists of the time. Thus, Bishop Isidore of Seville (560–636) and Rhabanus Maurus, Bishop of Mainz (776–856) produced works containing lists of known animals. These include some insects, among them lice, fleas and bed bugs, in a rough classification, but little more. Isidore, who was interested in etymology rather than entomology, notes that the louse was called *Pediculus,* because of its little feet (dimunitive of pes); the flea was called *Pulex* because it breeds in dust (pulvis). Neither he nor Rhabanus Maurus refer to Aristotle or even Pliny.

One of the last early mediaeval authorities worthy of mention is the Holy Hildegard,[131] Abbess of Rupertsberg near Bigen (1098–1179). She compiled a work which was partly a pharmacopoeia and partly a bestiary. Though she mainly wrote with practical ends in view, occasional passages of natural history intervene. Thus, she states that 'The Flea is warm and moist and grows from the dust of the earth. When the earth in winter is damp and inwardly warm, the flea springs to the earth and burrows therein. When, however, in summer the heat dries the earth, the fleas come out of the earth, spring on to men and annoy them.'

Arabian Interlude

The renewed acquaintance with classical knowledge came, in a roundabout way, from the Byzantine Empire. Creative science was equally dead in the eastern Mediterranean, but there was more activity in the preservation and copying of the works of antiquity. Many of the scholars involved were Syriac-speaking Nestorian heretics, who were persecuted by the Byzantines; they eventually migrated to Mesopotamia and later to southwest Persia. In their capital at Gondisapur, translations of many ancient texts were made. From the seventh century onward, the Arabs began to take over leadership in learning, and scholars from Gondisapur migrated to Damascus. The centuries from 900 to 1200 saw the highest development of Arabian science, which spread through the enlarged Moslem Empire. Contacts with west European scholars occurred in southern Europe; in Salerno, for example, and especially in southern Spain. As a result, some of the classical learning was translated from Arabic texts into Latin, from about 1000 onwards, by learned Jews (Avicebron, Maimonides and

Averroës) as well as some north European scholars. In addition to re-introducing classical science, the Arabs were responsible for some independent obervations, especially in chemistry and medicine.

An outstanding early Arabian zoologist was Al Gahiz (died 868). Unfortunately his works were never fully translated and many of his observations are known from later Arab writers (such as Al Qazwini, d. 1283, and especially, Damiri, 1349–91). From these various sources, Bodenheimer[27] quotes the following mainly from Damiri: *On fleas:* 'The flea is a creature with immense leaping powers. By God's grace, he leaps backwards to see who wants to catch him; otherwise he would easily be caught.' (Al Qazwini says 'he hops right and left until hidden'). 'According to Al Gahiz copulation lasts long; the eggs eventually hatch and from them the young are born. These live first in dusty dark corners. The flea is most troublesome in winter and early spring ... It has the appearance of an elephant, with dog-teeth to bite and a trunk for sucking blood. Eating it is forbidden, but killing it is allowed'. *On bed bugs:* 'They feed on warm blood and have a crazy preference for man. They have no protection, so one easily sees them. In Egypt and similar lands, they are very prevalent'. *On lice:* 'Thus says Gahiz: the nature of the louse is that, on red hair they are red, on black hair, black, on white, white; and if one changes the hair, they will change colour. The female is larger than the male ... Lice attack hens and pigeons and occur on apes ... Lice will infest any clothing except those of lepers.' *On scabies:* Perhaps the most significant original observation of Muslim scientists, in the field we are considering, relates to scabies. For the first time the disease is associated with the presence of mites in the skin, though the causal relationship was somewhat equivocal. I shall deal with their observations in a later chapter.

The Late Middle Ages: Scholasticism

Singer[288] suggests that the mediaeval period in Europe can be divided into two parts: the Dark Ages from 400 to about 1200, and a later Age of Arabian Influence, when the classical texts translated from Arabic into Latin began to circulate (as well as other Arabic learning). It was still too early for much original thinking, but scholars were compiling encyclopaedias of knowledge, including the work of classical times. Paradoxically, these scholars were all churchmen, notably the Dominicans, Albertus Magnus and Thomas Aquinas and the Franciscans Robert Grosseteste and Francis Bacon. Yet orthodox religion was still inimical to the heathen doctrines of classical philosophers and scientists and the biological works of Aristotle were banned by church synod several times in the thirteenth century.

Pre-eminent in the scholastic movement was Albert von Bollstädt, or Albertus Magnus (1193–1280) who, though he travelled about Europe, spent much of his life at Cologne. Despite the danger of being attainted heretic, he quoted Aristotle extensively, and even showed signs of original observation, in his great work *De Animalibus*. The original manuscript, in the Cologne state archives, dates from between 1255 and 1270; in later centuries printed copies circulated in Europe. Among the 450 species of animals mentioned there are 33 insects. The following is a rough translation of the sections on fleas and lice, from the 1495 Venice edition.[5]

Fleas originate in moist warm dust, if this has been in close contact with a warm animal body. They are black, round animals, which suck blood through their proboscis, leaving swellings at the sites. It has hind legs for jumping as well as six crawling legs. It moves very fast in proportion to its small body. It sucks a great deal of blood and eventually excretes it, black and dry. Its eggs, like those of lice, are lenticular. Small males are always found together with large females. The fleas born in March or April die in May. In that month very few fleas are found. Later some live into the winter and these are particularly unpleasant in August. The Louse is a worm which is generated from corruption in the ends of the pores of men and aggregates in animal heat or in the folds of garments, and similar forms are generated in animals, especially birds of prey. Furthermore, it seems that greedy boys have many lice which come from rich food, especially from figs, since the juice of unripe figs generates lice. It is said that lice have many legs, actually there are six. Their colour varies according to the nature of the humour, whose decomposition gave rise to them . . . There is another kind of louse which Galen calls the vulture louse or in Greek *memluke*, which lives in the human groin and grows among the hairs and under the armpit, whose bites cannot be seen unless a large number are bred . . . [presumably pubic lice].

Among the contemporaries of Albertus Magnus was a Franciscan known as Bartholomew Anglicus, or Bartholomew the Englishman[19] (fl. 1230–50) at one time professor of theology in Paris. He wrote a somewhat less ambitious encyclopaedia *De Proprietibus Rerum,* which was also subsequently printed in various countries. English editions appeared in 1495 and in 1535 and I quote the part of the section on lice and fleas from the latter.

De Pediculo (xviii.88). A Lowce is called Pediculus and is a worme of the skinne, and hath the name of pedibus, as Isido sayeth [Isidore of Seville, 570–636]. And greeveth more in the skinne and with creeping than he doth with biting and is gendered of right corrupt aire and vapours, that sweat out between the skinne and the fleshe by pores, as Constantine sayeth [Constantine the African, 1017–87]. Oft as he sayeth, lice and nits gender in the head or in the skinne . . . and expositours say that lice gender of sanguine humour and be red and great, and some of fleumatic humours and they be soft and white, and some of choleric humours and be citrine, long, swift and sharp; some of melancholy humour and they be coloured as ashes and slow in moving . . . And as there be divers kinds of beastes, so there be divers manner of lice, as it fareth in hogs, his louse is called Vsia . . .

De Pulice (xviii.89). The flea is a little worme and greeveth men most, and is called *Pulex,* and hath that name for it is namely fed with powder, as *Isidore* sayeth, libro. 12. And is a little worme of wonderfull lightnesse, and scapeth and voydeth perill with leaping, and not with running, and waxeth slowe, and fayleth in cold time, and in Summer time it waxeth nimble and swift. And though it be not accounted among beastes that be gendered and knowen among beasts by the medting of male and female, yet he multiply his own kinde by breeding of Neetes; for they breed certeine neets in themselves or comming of Neets and many Fleas do come of one Flea. And the flea is bred white and chaungeth colour, and desireth bloud and biteth, and he stingeth the flesh that hee sitteth on, and sucketh the thinnest parte of the humour that bee betweene the skinne and the flesh . . .

The accounts we have been considering were written by scholars interested primarily in knowledge for its own sake. But some early references to insects were made by physicians and pharmacologists concerned with them as pests or even as components of medicine. So far as these deserve mention, they would most appropriately be discussed in the medical history of ectoparasites (p. 176). But the various editions of the early German pharmacopoeia known as the *Hortus* (or *Ortus*) *sanitatis* (Garden of Health) tended to become more of a very simple, illustrated encyclopaedia. The earliest edition seems to be that of one Johannes Wonneck von Caub (or Cube), according to Bodenheimer,[27] about 1480. This became very popular and was printed in various forms illustrated with more or less primitive woodcuts. It was also plagiarized; for example, about 1520 a Dutchman called Van Doesborgh printed a close imitation in Dutch, under the title *Der Dieren Palleys* (The Palace of Animals). This, in turn, was translated into English by one Laurence Andrewe[149] (fl. 1510–37) then living in Calais, under the title *The Noble Lyfe and Natures of Man.* The following extracts are taken from the facsimile reproduction of Noel Hudson, who rediscovered this unusual book. 'Flees be bred or they growe out of filthy corners in houses and it is a litell blacke worme and it byteth for whait is warme or ayenst rayn and specyally more be nyght than be daye when one wyll take them they spring awaye.' 'A louse is a worme with many fete and it cometh out of the filthi and onclene skyne and oftentymes for faute of atendance they come out of the fleshe thrugh the skyne or swet holes'.

It will be very evident that these extracts show nothing more than a popularization of the great scholars two or three hundred years earlier. Yet the fourteenth to sixteenth centuries did, in fact, witness the operation of several factors which brought about the renaissance. Two of these have already crept into my account: the introduction of printing and the translation of works into national languages. We should not, however, assume that printed books were universally welcome; those who possessed libraries of beautiful manuscript books found them crude. Frederigo of Urino (1444–82), says Burckhardt,[41] would have been ashamed to own a printed book. Nevertheless, printing and (rather later) use of the vernacular, accelerated the spread of knowledge. In addition, as we are taught at school, the fall of Constantinople (1453) and invasion of Greece resulted in extensive westward migration of scholars. They brought with them some Greek texts of the classics, which corrected some errors and filled some gaps in the existing Latin texts which, as we have seen, were based on Arabic versions; and these, perhaps, on Syriac translations from Greek. As Raven[260] remarks, 'What the Greek New Testament did to mediaeval religion, that the Greek of Aristotle and Galen did to medieval science: it revealed the vast size of the accretions, and the startling contrast between the originals and their contemporary representatives. It was not that the fables were false to the facts but that they were false to their own sources.'

The Renaissance

That the renaissance flowered first in Italy is due to complex causes described so industriously by Burckhardt.[41] One of the reasons for the awakened interest in the classics in that country was the feeling of identity with the Romans and the Latin tongue. In the publication of these improved texts, the scholars of the time felt impelled to comment on them; and it was perhaps these commentaries which fostered the revival of original thought and led to the full blossoming of the renaissance. The new generations of scholars were often physicians rather than churchmen; but their interests ranged widely and they published works of rhetoric, poetry, natural history and even mathematics. A good example is provided by the north Italian Julius Caesar Scaliger (1484–1558) whose early history is uncertain (his autobiography differs from other less favourable accounts). Certainly an able soldier, he at some time studied medicine and for the last thirty years of his life published profusely. His scientific contributions were in the form of commentaries and criticism; for example, his commentary on Aristotle which forms a large folio book, with the classical texts in Greek and Latin side by side and his own comments underneath. As in theology, diverse interpretations of the ancients led to a series of polemics, well illustrated by Scaliger's vigorous controversy with a younger compatriot, Geronimo Cardanus[55] (1501–76). The latter also trained as a doctor; but he was unsuccessful at first and turned his talents to mathematics. Finally, however, he successfully re-entered the field of medicine and became successful and famous (though imprisoned for heresy towards the end of his life). Two of his books, caustically criticized by Scaliger, were *De Rerum Varietate* and *De Subtilitate Rerum,* which included classical references, accounts of experiments, and anecdotes. Scaliger's[277] comments were in the form of an erudite, though somewhat unreliable book. *Exercitationes de Subtilitate ad Hironymum Cardanum.* His style is vigorous and sarcastic and, though showing signs of original judgement, he seems more interested in refuting Cardanus than in arriving at the truth.

One of the most interesting subjects of dispute among these sixteenth century scholars of natural history was that of spontaneous generation, in which the authority of Aristotle came into conflict with the observations which naturalists were beginning to make for themselves. In earlier times, such inconsistencies were noted, but did not seem to have aroused much interest or discussion. Thus, I have mentioned the statement of Albertus Magnus that, among fleas, small males were found together with large females; whereas Bartholomew Anglicus[19] said that fleas were not among the animals with different sexes which mated, although they did, nevertheless, produce fertile eggs.

Scaliger, however, brings the matter into his altercation with Cardanus. The latter had suggested (*De Rerum Varietate* VII.28) that animals could be generated from putrefaction; and thereupon fed on the putrefied material. Scaliger rejects this, but for an odd reason (*Exer-*

citationes CXC.2). He cites the ancient belief that bees are generated from animal carcasses (referred to, indirectly, in the Bible: 'Out of the strong came forth sweetness' (Judges 14:14). Also, in the *Hortus Sanitatis*,[149] there is a crude wood-cut of bees emerging from a dead ox. It is possible that drone flies (*Eristalis tenax*) which breed in ordure, were mistaken for bees).[27] Scaliger remarks: 'But, if from a dead calf, came bees, and from bees, worms and something else from these, how is the cycle completed to return to the calf?'

He does not really get to grips with the problem, and he is easily led into anecdotal byways. Thus, Cardanus had elsewhere suggested (*De Subtilitate Rerum* IX) that Carthusian monks were not plagued by bed bugs because they do not eat meat. He is not, however, categorically sure and suggests that some other cause, such as cleanliness, might be responsible. Scaliger finds this an excuse for some rather heavy-handed sarcasm (*Exercitationes* CCXLVI.5).

Tales about Carthusian bugs have pleased you, some nonsense about them was included in (your book). You add to these lies matter no less false; that they (the Carthusians) are not attacked by them because they abstain from meat. Would that this had been known to Pythagoras! [a vegetarian] Do you not remember that dogs are not sought after by bugs and fleas avoid horses? Mice, however, breed so many fleas that, unless carefully examined, they seem to be covered by a skin of fleas. So be it. Suppose that no Carthusian is bitten by bugs. The question remains, can the bugs breed in their bed-chambers? Now, the beds in Toulouse do not eat flesh, yet they are notorious for this pest! . . . They flourish particularly in couches of firwood, especially when the straw has grown old. They do not flourish if the straw is sewn up in a cover. They are also bred in parchment books, and, in marvellous profusion, in hen coops . . . The Scots dye their clothes with saffron against lice . . . In the island of St Thomas, abundant lice afflict the negroes; but our people are not troubled. This year, no flea affected a woman suffering from jaundice; once she had recovered, they invaded her as before. The cause, I judge to be the bitterness of the yellow pigment. Everybody knows how lice flourish in the public hospitals . . . As soon as any of the wretched people die, immediately the lice leave their former lodging, that is, the corpse.

A third medical naturalist of the time was the Englishman, Edward Wotton (1492–1555) who studied at Oxford and for three years at a medical school in Padua. Later a fellow, and eventually a president of the Royal College of Practitioners, it was his book of natural history *De Differentiis Animalium*[326] which won him international reputation. Wooton, however, devoted himself to zoology, explaining in his dedication that plants had been well covered by the work of the French botanist Jean Ruel, whose fine book *De Natura Stirpium* has appeared in 1536, and minerals by *De Natura Fossilium* of the German, George Bauer. Of the ten books of *De Differentiis Animalium*, the first three are general, IV deals with man, V with mammals, VI with reptiles, VII with birds, VIII with fishes, IX with insects and X with molluscs and crustacea. The insects, which included other land arthropods, were roughly classified by the presence of wings and legs. Fleas, lice, bugs, ticks and itch mites were grouped together (IX.292) as 'Those which feed on the humours of living flesh'. These, he says, 'produce what are called "Lendes" from which nothing is born . . . but they come forth from persistent dirtiness'. The section on lice closely follows Aristotle and includes his remarks about their emerging from little dry boils, a mention of the lousy disease which killed Alcaman and Pherekydes, the tale of lice engendered in wool from a sheep killed by wolves and so forth. Also there are notes about lice of other animals; and this leads him to discuss ticks. Then he turns to lice of fish and marine fleas and hence to true fleas. The short discussion on the latter is no improvement on Albertus Magnus. Finally, there is a page on bed bugs, which is mainly concerned with their use in medicine, from Dioscorides and Galen. His own brief comments at the end of the chapter are more etymological than entomological.

These remarks bear out the opinion of the great Swiss zoologist Gesner that Wootton 'teaches nothing new but gives a complete digest of previous works on the subject.' Raven remarks, 'There is no evidence that he specially cared for, or indeed had ever seen, any of the creatures which form the subject of his quotations. He had obviously read widely, copied his extracts, and fitted them together diligently.' The book was 'admirably printed in 1551 at Paris, by Michael Vascosanus . . . one of the first printers to give up the black letter type.'

In the early seventeenth century were published the two earliest substantial books on entomology: Aldrovandi's *De Animalibus Insectis*[6] (1602) and Mouffet's *Insectorum sive minimorum animalium Theatrum*[219] (1634). Both of them had their roots in the previous century, as we shall see.

Aldrovandi's work was the fruition of some 50 years' study. Born of a noble family in Bologna in 1522 he was a lively boy and twice left home, once on a year's pilgrimage to Spain. On return, he began to study eagerly but later showed signs of heresy which led to his interrogation by the Inquisition in Rome. Released, he pursued his studies and became a doctor of medicine and philosophy at the age of 31; and seven years later was appointed a professor. In 1568, he helped to found, and later directed, a botanical garden in Bologna. Subsequently, his new ideas on pharmacology aroused the enmity of the local doctors; but he was supported by the Pope and lived to a famous and prolific old age. His special virtue was his interest in a natural history collection which became a substantial museum. *De Animalibus Insectis* is not strikingly original and the system of classification is based on the same simple items as that of Wootton. But Aldrovandi attempts to simplify this by a key, depending on alternatives (land or water insects; with or without wings; or feet, etc.). Furthermore, many more kinds of insect are mentioned and there is a far more complete survey of earlier works on each. Thus, the subject of ectoparasites with which we are concerned, occupies only two paragraphs (say 450 words) of Albertus Magnus' work and perhaps 1000 words of Edward Wotton's book; but Aldrovandi devotes to them no less than 19 folio pages (say, 10 000 words). Only a fraction of this could be deemed natural history. After a short introduction to each pest, he considers synonyms (in various languages) and also doubtful references. Then a section of the genus and its form and one on its place of generation. Methods of control and uses in medicine receive attention and, finally, relevant proverbs and epigrams in classical literature, some of which I have mentioned earlier. On the credit side, his text is sprinkled with useful marginal headings of minor subjects, like those of many contemporaries; but not all are so well supplied with references for their statements. These authorities include all those I have noted, from Aristotle to Cardanus and Scaliger, as well as other more medical authors, such as Galen, Mercurialis and Avicenna. On the other hand, his chapters tend to be disordered and frequently repetitive; one feels that the text would be vastly improved by trenchant editing.

In the beginning of the work, Aldrovandi deals with the anatomy, life history, physiology and reproduction of insects in general terms. On the whole, he is conservative; thus he follows Aristotle in considering that, being cold blooded, they do not need to cool themselves by breathing. Accordingly he rejects Pliny's cautious speculations on the subject. Since they have no lungs, they cannot breathe, he considers; this is evident from the fact that a fly, immersed in water for a long time, is not drowned. Similarly, he follows Aristotle's account of the various modes of generation of insects (p. 127). In later sections he manages to reconcile the master's statements with observation in various ways. Thus, we have seen earlier (p. 97) that he accounts for bug infestations in the neglected beds of the poor, because accumulations of human dirt tend to breed bugs, as Aristotle says. Later he refers to an observation of Cardanus that bugs are generated in old books, and he speculates on the possibility that the linen in the paper could perhaps have had contact with the human body; but on the whole he is sceptical. To conclude the section on bugs, he mentions a range of plant-feeding bugs and provides some rather crude figures. Elsewhere in his book, he includes some quite passable woodcuts of butterflies and other insects (though none of ectoparasites). These were based on quite excellent water colour drawings made for him by contemporary artists.

On the subject of lice, Aldrovandi has to cope with the varieties mentioned by classical authorities. To begin with, Aristotle had mentioned ordinary and 'wild' lice. The ordinary louse causes no problem, though so far as I can see, he does not differentiate between the head and body variations. As regards the 'wild' louse, he notes that Scaliger (*Exercitationem* xcv) has apparently confused it with a tick ('*Ricinum hominis*'). This is the creature which Albertus Magnus mentions as infesting pubic hair, axillae and eyelashes (and, following Galen, calls the 'Vulture louse'). Aldrovandi fairly certainly knows that the crab louse is meant, since he gives various vernacular synonyms, including the French 'Morpiones' (modern French, morpions). In the same section, itch mites are considered as a form of louse, though he mentions Aristotle's name 'Acarus' and the vernacular Italian 'Cyrone' and 'Pedicellus'. He quotes Scaliger's remarks (taken from Arabian sources) about removing the creature from its burrow, cracking it on the fingernail, etc. so that it is clear that the scabies mite is being described.

Concerning the generation of lice, Aldrovandi quotes the sentence of Aristotle about lice emerging from dry boils in the skin. The question then arises, whether the lice in the flesh are generated from 'corrupt' blood, as suggested by Theophrastus, Aristotle's pupil. This leads to some complicated yet ingenious arguments. If lice are generated from blood, surely they should be reddish in colour? But this is not so; and somewhat later, he writes (like Bartholomeus Anglicus) of lice deriving their various colours from different humours. On the other hand, some authorities, such as Avenzoa (following Galen) believe lice to be generated between the flesh and the skin. But, says Aldrovandi, as everyone knows, lice in boys are most commonly produced in the scalp, where there is but little flesh! He seems, finally, to conclude that lice, whether generated inside or outside the skin, are produced from the interaction ('*concoctionis*) of excretion, sweat and dirt and also, apparently, by contagion with lousiness. Later, he points out that soldiers in camps are often lousy (and their officers too!) from not being able to change their clothing often enough. Likewise, poor folk suffer from lice for the same reason. Crab lice are likely to infest those who consort with harlots.

After these observations which, one would suppose, would bring him away from the mediaeval concept of spontaneous generation, he reverts to an abstruse discussion of Avicenna's two kinds of heat: 'natural' and 'universal'. The former does not create life, but the latter does, through the medium of fermentation (*concocto*) or putrefaction; and this is the sort responsible for generating lice.

In considering the generation of fleas, Aldrovandi begins with the familiar quotations from Aristotle about bugs, lice and fleas producing the non-viable 'lendes', but he also mentions Bartholomew who stoutly maintained that the flea's eggs do, in fact, give rise to more fleas. Aldrovandi then seems to agree with this (*Ego quoque Lendes nihil gignat, non puto . . .*) but hastens to add that they do, of course, arise in warm dust as Aristotle has said. Finally, however, Aldrovandi returns to the attack in a final chapter (7), *De Lendibus*. Once again he quotes Aristotle but continues that we cannot believe this, since many lice sometimes develop from a few, and furthermore, common people often are anxious to get rid of the nits (lendes) rather than the lice (and he gives some remedies for this purpose).

I have already mentioned that the *Insectorum Theatrum*, published in 1634, was written in the previous century. But, unlike Aldrovandi's book, it was compiled from the work of several learned men. It seems to have been begun by Thomas Penny (1530–88), who studied divinity and medicine at Cambridge, remaining there as a Fellow for some years after graduation. As a preacher, he strongly supported the Reformation; apparently with insufficient reverence for the establishment, for he attracted the censure of an Archbishop. Whereupon he decided to devote himself more completey to medicine and travelled abroad for several years (1665–9). He began with a visit to Conrad Gesner, in Zurich, from whom he may have learnt much, though the exchange of information was not all one way. Penny had already acquired considerable knowledge of plants and his drawings and descriptions were accepted by Gesner, as well as other European naturalists. Fairly soon after their meeting, Gesner died (of bubonic plague) at the age of 49. Penny may have been present at his death and he afterwards assisted in sorting out papers Gesner left behind. Some of the entomological material may have formed the embryo of the *Insectorum Theatrum*.

In later life, Penny acquired an M.D. and practised medicine in London, becoming a Fellow of the College of Physicians in 1582. He did not, however, lose his interest in natural history and found time to botanize in various parts of England. Also, over a period of 15 years, he collected entomological information; partly from older authorities, partly from his own observations and also by the help of various correspondents. Among those, acknowledged in the *Theatrum*, were eminent foreigners such as the Quickelberg family in Amsterdam, the French botanist De l'Ecluse and J. Camerarius, a doctor and botanist of Nuremberg. The British authorities include Sir Thomas Knivet and his uncle Edmund (members of an erudite Norfolk family), Peter Turner (son of the more famous William Turner), William Brewer, John James, Roger Brown and Timothy Bright. It is evident, then, that a considerable number of original records were added to the basic traditional learning which, in the Theatrum, was mainly derived from Wotton and Gesner.

After Penny's death he left his collection of papers to his young friend Thomas Mouffet (also spelt Moffet and Muffet). The son of a Scottish haberdasher living in London, Mouffet was born in 1553 and educated at Cambridge. Later, he too studied in Switzerland and graduated M.D. in 1578. After further travels abroad, he returned to England and practised medicine; first in Ipswich and then in London. In his maturity he attended nobility (Duchess of Somerset, Sir Francis Walsingham); and he met such famous people as Sir Francis Drake and (on a visit to Denmark) Tycho Brahe. Travelled, learned and well connected, he was towards the end of his life, patronized by the Earl of Pembroke and his wife, who persuaded him to settle at Wilton (near Salisbury) on a pension; he was even elected to parliament. He was married twice; first at the age of 27, and 20 years later, to a widow with children. His own daughter Patience has sometimes been identified with the Little Miss Muffet of the nursery rhyme, perhaps because of her father's interest in spiders!

Mouffet sorted and arranged the papers he inherited from Penny, removing a large number of what he considered to be irrelevancies; he added a little additional material and a large number of figures. Raven, who made a careful study of the book, suggests that Penny was more of a naturalist than Mouffet, who had, however, 'a flair for a telling phrase, a literary gift which is wholly unlike the dry scientific precision of Penny's style . . . Moreover, he is incurably medical and loves to accumulate masses of lore and learning scarcely relevant to his theme, and reminding us of his books on diagnosis and dietetics'. At all events, he completed the *Insectorum Theatrum* in 1590 and wrote a dedication to Queen Elizabeth. A fine title page was prepared, with portraits of Mouffet, Penny, Gesner and Wotton. But long delays followed and it was still unpublished in 1603 when the queen died; and Mouffet died the following year.

The *Insectorum* papers passed to Mouffet's heir, who was too poor (or too indifferent) to ensure their publication. A dedication to James I was written, but without succes. Finally one Darnell, the late Mouffet's Apothecary, offered the manuscript to Sir Theodore Mayern, an eminent French doctor who had emigrated to England about 1606 and became physician to James I. Sir Theodore had diverse scientific interests outside medicine and it was he who finally published the *Insectorum Theatrum* in 1636, with a dedication to Sir William Paddy the King's chief Physician. A new frontispiece was used, devoid of the original rococo ornament and depicting the insects rather than the authors. Even this, however, is not the end of the story. In 1607, a scholarly English cleric named Edward Topsell (1572–1625) had published a book on the *History of Four-footed Beasts*. It was not particularly original, being largely based on Gesner. This book was republished in 1658 together with an English translation of the *Insectorum Theatrum*, edited by one John Rowland, M. D. It is from this edition that I will quote anon.

A comparison of the *Theatrum* and Aldrovandi's book suggests that their texts were independent, though some similarities appear, due to both of the authors quoting extensively from earlier authorities. They use roughly similar schemes of classification, though Aldrovandi begins with land- or water-insects, whereas Mouffet starts with winged or wingless forms. On the whole, Aldrovandi's scheme is more sophisticated, since it divides the winged group into those with two or four wings and takes note of those with membranous, horny or scaled wings. On the other hand, the actual names in the *Theatrum* are important, because many of them were adopted by Linnaeus.

Bodenheimer[27] devotes some space to comparing the two books. He analyses the species described in each (so far as this is possible) and assigns them to families. It seems that Aldrovandi mentioned 83 species (70 genuine), whereas Mouffet mentions 64 (63 genuine). The former author's fauna is mainly that of northern Italy, whereas Mouffet combines about two thirds of English species with various exotic forms supplied by foreign correspondents. Bodenheimer tends to think Aldrovandi underestimated; but Raven in a later and detailed comparison, is quite definitely of the opposite opinion. The *Insectorum Theatrum* devotes about the same amount of space to human ectoparasites as Aldrovandi; that is, about 10 000 words. Marginal headings occur, but are fewer, and references are incorporated in the text.

Chapter 22, *Of the six-footed Worms of living Creatures, and first of Lice in Men,* begins with historical references and synonyms; and there is a tolerable figure, considering that it was probably drawn without the help of a lens. There is no doubt about the identity of 'tame' and 'wild' lice:

... those the English call *Lice,* and these *Crablice* ... The tame ones breed of corrupt bloud and are lesser and reddish, from Fleame white, from melancholy and adult humours black and from mixt humours they are of divers colours, as *Petrus Gregorius* noted I.33. If you rub them gently between your fingers, you shall see them foursquare and somewhat harder than Fleas, whence in the dark you may easily find the difference. These that breed in the head are bigger, longer, blacker and swifter, those that breed on the body are fatter, bigger bellies, slower, darkish white and marked with blackish streaks.

Somewhat later, after a diversion about Diodorus' story of the locust eaters and flying lice which emerge to destroy them (p. 104) he returns to the vexed question of crab lice.

Agatharcides speaks of these lice, but he saith they are like unto Ticks. They chiefly fasten on the chin, eyebrowes, and the privities full of hair, the groin and the arm-pits, their body is more compact, their nib is sharper, they bite more and tickle less. For Tykes will sometimes enter deep in the skin with their nose, that you can hardly pull them out but with the loss of their heads, and they seldom wander, but they bite cruelly, and make themselves a hollow place, and there they stand fast. Some call these Lice in Latine, Cicci, some mens Tikes, other Vultures lice; Aristotle calls them wilde Lice, *Hist. Animal,* 1.5, c.31. it is harder than a tame Lowse, and is more hardly removed from the place it bites.

The long section on the generation of lice comes very close to

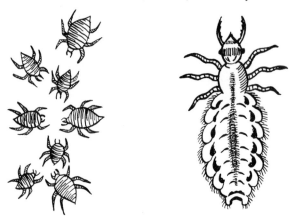

Fig. 6.1. Two figures copied from the *Insectorum Theatrum*.[219] *Left*, Wall lice (bed bugs); *right*, a louse.

Aldrovandi, with the same discussion of the relative merits of 'flesh corrupted' and 'putrefied bloud'. Mouffet, however, states that in children's heads, red lice are sometimes found; and also on people recovering from a 'putrid Synochus' (whatever that may be). As alternatives to corrupt flesh or 'bloud', ancient medical authorities are quoted for breeding from '. . . hot excrements of the second and third concoction putrified, not sharp, nor bad.' To understand this, says Mouffet

. . . we must know that when bloud is changed into the substance of the Limbs, many kinds of excrements are produced, whereof some are dissolved by insensible transpiration, others by sweat, others turn to filth, others stay in the skin: Those that are retained in the upper skin, make dandruff, if they stay in the depths of the skin, or are bad and sharp, they cause sore heads. But since I have observed that in some that were in a Consumption incurable, where the sharpness of the humour eats up the very roots of the hairs, Lice come forth abundantly, why may I not think by their leave that they may breed at first from sharp humours?

A faint clue from a recent authority leads, alas, into yet more medieval metaphysics. '*Scaliger* would prove that Lice breed not from putrid humours, because herbs grow from the seed without putrefaction . . . But by his leave I must say that *Scaliger* or the Apostle must be mistaken: For so *St Paul* (1 Cor. 15) *That which thou sowest is not quickened unless it die.* But if death is a corruption, as the Philosophers say, then *Scaliger* was deceived, and (yet keeping the Laws of friendship) we may deservedly rejected his opinion.'

Towards the end of this section, Mouffet refers to the legend about lice breeding in woollen clothes made from a sheep killed by a wolf. He continues: '. . . which grant that it be an invention of *Aristotle* and *Pliny*, yet experience teaches us that cloathes smeered with Horses grease, will breed Lice presently.' What experience, one wonders?

Mouffet was, of course, familiar with louse eggs; but he deals with them quite separately in a later chapter (35) *Of Nits*.

Trotula not improperly calls them . . . hair eaters, for as Snails live on the juice of herbs, so these live on the moysture of the hairs and feed thereon. The philosopher affirms that they proceed from the copulation of Lice, and therefore are called their eggs. They are like the flowers of Jesamine, that grows with us. For as Jesamine brings flowers without seed, so lice bring forth eggs without young ones in them.

Mouffet is ahead of Aldrovandi in recognizing that itch mites are not merely a variety of lice, nevertheless certain confusions are still present. He devotes a separate Chapter (24) on the subject *Of little Lice called Syrones, Acari, and Tineae, or Hand-worms, or Mites in living creatures*. Though there are quite a number of other sixteenth century references to itch mites, Mouffet's compilation is by far the most comprehensive. I discuss it more fully in a later section of the book (p. 208). The *Theatrum* is reasonably sound in regard to actual scabies mites; but the chapter includes very dubious accounts of mites in the eyes and teeth. In addition, there are possible allusions to tinea of the scalp (a fungus disease of children), warble fly maggots on the backs of cattle, and chigger fleas that penetrate the skin in certain tropical countries.

On the subject of the generation of bugs, Mouffet shows evidence of originality.

They are bred, after Aristotle's opinion, from the moisture the sweats forth on the body of living creatures . . . but without doubt they arise from other humours corrupting about beds that sweat out of wood be degrees. Also they propagate by copulation, as *Pennius* observed at *Orleance;* for whilst he kept company with a Spaniard born at *Capera,* he strove to draw his sword to cut off a bough; but when he could hardly do it for the rust he was forced to cut the scabbard where he found abundance of Wall Lice with a great company of young ones and a multitude of whitish eggs of a watry complexion.

Somewhat later occurs the first record of bed bugs in England.

In the year 1503 when *Pennius* writ this, he was called in great haste to a little village called *Moreclack* near the Thames, to visit two Noblemen, who were much frightened by perceiving the prints of Wall-lice, and were in doubt of I know not what contageon. But when the matter was known, and the Wall-lice were catched, he laughed them out of all fear.

There appear to be two errors in this. In the first place, Penny seems to have been born about 1530; certainly he only graduated in 1551, so that it is unlikely that he was much of an authority in 1503! Inspection of the Latin version of the *Theatrum* reveals bad printing of the third figure of the date; but close examination suggests that it should be 1583. Again, the Latin version gives the place as 'Mortlacum'; presumably Mortlake. In the next Chapter *Of the Flea,* Mouffet observes,

... their hinder legs are bent backwards toward their bellies and their forelegs toward their breasts, as four-footed beasts are, as it is usual almost in all Insects to whom Nature hath given but four feet ... The ends of their feet are divided into two parts and are hooked and sharp ... The point of his nib is something hard that he may make it enter the better. It must necessarily be hollow, that he may suck out the blood and carry it in. They seek out the most tender places, and will not attempt the harder places with the nibble; with two very small foreyards that spring out of their foreheads, they both prove their way, and judge of the nature of an object, whether it be hard or soft; where they bite they leave a red spot as a Trophie of their force, which they set up.

This is quite impressive for naked eye observation: though it is somewhat curious that inspection which could reveal the claws on a flea's foot and the small antennae, could overlook the third pair of legs.

On the subject of the generation of fleas, Mouffet mixes fact and fancy. For example: '*Martyr* the Author of the *Decads of Navigation,* writes, that in *Perienna* a Countrey of the *Indies,* the drops of sweat that fall from their slaves bodies will presently turn to fleas.' A little later, however, we have:

They copulate, the male ascending upon the back of the female as flies doe, and they both goe, leap and rest together. They stick long together and are hardly pulled asunder. After copulation presently almost, the female full of Egges seems fatte; which though in her belly they seem long, very small, very many and white, yet when they are layd, they turn presently black and turn into little fleas, if we may grant what *Pennius* saith, that bite most cruelly ... Aristotle thinks that from them, be they Egges, Nits or little Wormes, no other creature breeds, and I should willingly subscribe to him, but that I think Nature made nothing in vain.

The Seventeenth-Century Watershed

In the course of the seventeenth century, some radical changes occurred in scientific thought. In the subject of biology, learned men gradually freed themselves from the apron strings of Aristotle; and this can be illustrated most appositely by the changes in entomology and, especially, the abandonment of the idea of spontaneous generation of insects.

To illustrate the advanced thought of the early decades, who better than Francis Bacon (1561–1626) and Rene Descartes (1596–1650). Both, of course, are more renowned as philosophers than as scientists; and the passages I shall quote were somewhat marginal to their main works. Nevertheless, it is extremely unlikely that anyone could have challenged their opinions on spontaneous generation at that time.

Bacon's remarks on the subject occur in *Sylva Sylvarum*,[16] a miscellaneous and apparently haphazard collection of observations on natural history phenomena, published in 1627, a year after his death. The relevant passage (cent. VII, 696) seems to me to be redolent of mediaeval metaphysics: 'So the *Nature of Vivifaction*, is best enquired of *Creatures* bred of *Putrefaction*. The *Contemplation* whereof hath many *Excellent Fruits*. First, in Disclosing the *Originall* of *Vivefaction*. Secondly, in Disclosing the *Originall* of *Figuration*. Thirdly, in *Disclosing* many *Things* in *Nature* of *Perfect Cretaures*, which in them *lye* more hideen. And Fourthly, in Traducing by way of *Operation*, some Observations in the Insecta, to work *Effects* on *Perfect Creatures*. Note that the word *Insecta* agreeth not with the Matter, but we ever use it for Brevities sake, intending by it *Creatures bred* of *Putrefaction*.

The Insecta are found to breed out of several Matters: Some breed of *Mud* or *Dung* . . . etc. . . . It is observed also that Cimices are found in the *Holes* of *Bed-Sides*. Some breed in the *Haire* of *Living Creatures; as Lice*, and *Tikes;* which are bred by the *Sweat* close kept, and arefied by the *Haire*, . . . *Fleas* breed Principally of *Straw* or *Mats*, where there hath been a little *Moisture;* or the *Chamber* and *Bed-straw* kept close and Not Aired . . .'

Descartes, an even more eminent philosopher, made distinct contributions to mathematical theory. But his biological lucubrations seem to me to be more in the way of illustrations of a philosophical system than studies pursued for their own sake. He deals with spontaneous generation in an isolated work *'Opuscula Posthuma: by R. Descartes'* published in Amsterdam in 1701. C. A. and P. Tannery considered it genuine and included it in their edition of his Works[74] (p. 499). It begins: *'Duplex considerada est Generatio, una sine semine vel matrice, alia ex semine.'* After a digression, he continues:

Every animal that develops without a womb, requires for generation, two principles stimulated into movement by Warmth; so that what may be called spiritual humour (the more delicate component) arises with vital humour (from the stronger component). When these two principles fuse together the heart awakes to life . . . Since so little is essential for the formation of an animal, we need not wonder that so many animals, worms and insects originate in filth.

If these two brilliant men could not question the generally accepted idea of spontaneous generation, it is little wonder that it coloured the ideas of that slightly eccentric, self-taught philosopher Thomas Tryon (1634–1703). In 1682, he published *A Treatise of Cleanliness in Meats and Drinks, of the Preparation of Food, the Excellency of Good Airs, and the Benefits of Clean Sweet Beds.*[304] He remarks of bed bugs:

Heat and Moisture is the Root of all Putrefaction: and therefore Bugs are bred in the Summer: but they live all the Winter, though are not then so troublesome. They harbour in Bedsteads, Holes and Hangings, Nitting and breeding as Lice do in Clothes. But all Men know that Woollen and Linen are not the element of Lice, but they are bred from the fulsome Scents and Excrements that are breathed forth from the Body. The very same Radix hath Bugs; and if there be any difference, they are from a higher Putrefaction and therefore a more noisome stinking Creature.

We have once again iterated the idea that 'Excrements and Breathings of the Body will generate Vermin' and 'If there was but a tenth part of the Care taken to keep Beds clean and sweet, as there is of clothing and Furniture, then there would be no Matter for the getting of Diseases, nor the Generation of Bugs . . .' I cannot but applaud his principles of good housekeeping as a husband, however I decry his biological principles as a scientist; and even with these, I must sympathise, since at that time neither ancient nor medieval nor renaissance scientific writers doubted that spontaneous generation took place on occasion. Corpses were said to 'breed' worms, dirt to 'breed' vermin, sour wine to 'breed' vinegar eels and so forth. Spontaneous generation is often fathered on Aristotle and is in his writings, but it was not so much a doctrine as a universal assumption.[288]

To overcome this obstacle, a firm belief in independent observation was required. One with such convictions was William Harvey (1578–1657), well known for establishing the fact of blood circulation. Harvey was also interested in animal reproduction which was the subject of his last published work, *Exercitationes de generationem animalium,*[116] 1651. An English translation appeared two years later, from the Preface of which I quote: '. . . the Reader . . . shall also understand, how unsafe and degenerate a thing it is to be tutored by other men's commentaries, without making a tryal of the things themselves: especially since *Nature's Book* is so *open* and *legible.'* Later, comes an important passage:

> ... all *Animals* may in some sort be said to be born out of *Eggs,* and in some sort out of Seed: besides, they are stiled *Oviparous, Viviparous,* or *Verminparous* rather from the issues themselves bring forth, than from the *original matter of which themselves were made:* namely because, they produce an Egg, a *Worm* or a *living creature.* Some are also said to be *sponte nascente,* creatures born of their own accord, not because they quicken out of putrid matter, but because they are begotten by chance, by nature's own accord, and by an aequivocal generation (as they call it) and by parents of a different species from themselves.

Though the last sentence is slightly equivocal, yet the general principle enunciated at first became widely known and quoted as *Omne vivum ex ovo.* Scepticism, however, grew slowly. Thus, John Johnstone (1603–75) of a Scottish family settled in Poland, produced a small but dense encyclopaedia,[161] with biological sections. He challenged the Aristotleian concept of generation of butterflies from dew, for he knew that they mated and laid eggs. On the subject of lice, however, he adopts the same view as Mouffet, rejecting their generation from blood or flesh, but accepting their origin in putrid humours given off from the skin, under the influence of warmth and humidity.

A categorical statement in regard to lice was made by Johannes Sperling in his posthumous work *Zoologica Physica*[293] (1661). Born in Thuringia in 1603, Sperling studied theology, but turned to medicine, and eventually became Professor of Natural Science at Wottenburg in 1634. According to Bodenheimer, Zoologica Physica was written as a series of lecture notes, the various chapters being divided into sections: *Praecepta, Questiones, Axiomata.* The entomological portion is appended to the larger portion on higher animals. One of the early, general Questiones deals with the problem of insect respiration; and on this subject, he adds little to Aristotle and Pliny. But on the generation of insect ectoparasites he has something new to say. On lice (X, 465), *Quaenam generatio?,* he states,

> they breed not from filth, not from excrement, not from flesh, not from blood, but from coition by the medium of nits, their eggs. Aristotle concedes that lice generate nits but states that nothing comes from them. But experiments disprove this idea with the demonstration that fertile nits turn into Lice.

The remarkable metamorphosis of entomology during the

seventeenth century was brought about by three factors. First, as we have seen with Harvey and Sperling, a growing confidence in one's own observation, as opposed to mere perusal of ancient books. Then, and no doubt proceeding from it, there was a desire to experiment; an inclination which seems to me quite natural to any curious person, however ignorant of modern scientific method. The third factor was the development of magnifying lenses, and especially their combination into microscopes.

It is not clear who invented the compound microscope, but it was known that Gallileo was using one in 1610, and one, apparently made by Zacharias Janssen in Holland, was exhibited in London in 1619. Insects were among the first objects to be scrutinized and, as early as 1619, the Italian Francesco Stelluti a colleague of Gallileo, was using a microscope to study insects and depict their external structure. It seems unlikely that Penny or Mouffet used a magnifying glass; but Theodore Mayerne certainly did, for he mentions it in the introduction to the *Theatrum*[219] and adds observations beyond those in the text.

And moreover, if you will take lenticular optick Glasses or crystal (for though you have Lynx his eyes, these are necessary in searching after Atoms), you will admire the dark red colours of the Fleas, that are curasheers [? cuirassiers]. and their back stiff with bristles . . . You shall see the eyes of the Lice sticking forth, and their horns, their body crannied all over, their white substance diaphonous and through that, the motions of their blood and heart . . .

As soon as microscopes came into use, the most ordinary things (needle-points, razors, cloth) were found to be of fascinating interesting, and common body parasites were among the earliest objects examined. Pierre Borel[32] (*c.* 1620–89) a French court physician, published an account of his observations with telescopes and microscopes in 1651. One hundred objects viewed by microscope are described briefly (from a sentence to a paragraph) and a few very simple figures provided. The 'horrid transparent louse' allows the blood circulation in its heart to be seen, he says; and he describes a battle with a flea as being like two-monsters in an arena. A flea's shape is like a lobster and the eggs can be dissected out from a pregnant one. Small creatures are found in scabies lesions, he says, but also claims that these occur in other skin diseases, including smallpox.

The wonder and excitement of the new world of minute creatures, evoked poetry of a sort from an early English microscopist, the physician and naturalist Henry Power[251] (1623–68):

> Of all th'Inventions none there is surpasses
> the Noble Florentines Dioptrick-glasses
> For what a better, fitter guift Could bee
> in this World's Aged Luciosity.

And soon he is celebrating the wonderful appearance of the common louse and flea:

> View but the wonders in a Louse through whose
> translucent Body all Her entrayles shows
> As through a Chrystall Watch I have often seene
> the Counter-motions of the wheeles within.
> There you may see the pulsing Heart driue on
> the Blood in its due Circulation.
> If Nature's skill in opticks you would trye
> then take a Flea and looks me at her eye
> Where in an Emarauld Iris shee hath sett
> (empaled with white) a pupill all of jett.

Power's poem recounts in verse his observations, which were recorded somewhat more soberly in his one book, *Experimental Philosophy (in three books) containing New Experiments: Microscopical, Mercurial, Magnetical (1664 or-3)*. The microscopical section would seem rather less 'experimental' than the other two portions, which involve filling tubes with mercury, manipulating loadstones and so forth; and yet the same spirit of discovery is involved.

In the mid-seventeenth century this spirit was infusing a number of talented Englishmen, and their mutual interests led to the foundation of the Royal Society, of which Power was an early member. His slight, but ecstatic contribution to microscopy was soon eclipsed by the greater achievements of Robert Hooke. Born in 1635, Hooke graduated at Oxford, where he assisted Boyle in his experiments with gases. At the age of 27, he was appointed curator of experiments to the Royal Society and remained associated with it for the rest of his life. He contributed to diverse sciences, notably physics and astronomy, and had relations with various famous people, which were sometimes marred by acrimonious disputes. His versatile and ingenious mind forestalled several important discoveries, though often failing complete proof. After the fire of London, he assisted Wren in the re-building plans.

Hooke's microscopical observations (together with accounts of experiments on refraction, interference colours and barometric pressure) were published in 1665 in *Micrographia*.[143] Samuel Pepys[238] bought an early copy, recording on 20 January '... so to my

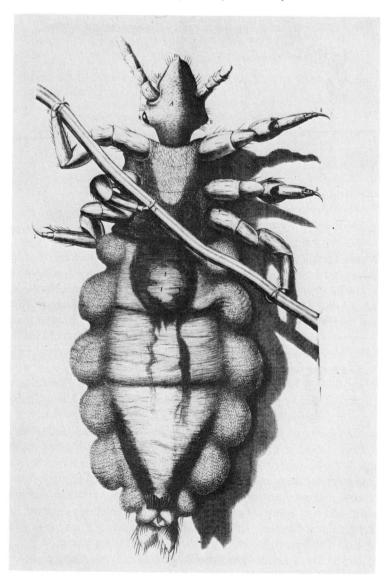

Fig. 6.2. Robert Hooke's picture of a louse in *Micrographia*,[143] 1665. By courtesy of The Wellcome Trustees.

booksellers and there took home Hooke's book on microscopy, a most excellent piece, and of which I am very proud.' Next day he sat up reading it until after 2 a.m., stating that it was '. . . the most ingenious book that ever I read in my life.' In the chapters on the Flea and the Louse, Hooke is essentially concerned with their external appearance. He notes (as Power did) the peristaltic movement of the gut of the louse and also the ramifying tracheae, which he wrongly assumed to be blood vessels filled with white blood. But the excellence of the large and detailed figures is astonishing; possibly they were re-drawn by a professional engraver (Figure 6.2). Elsewhere in *Micrographia,* Hooke discusses spontaneous generation, in relation to gall insects and gnats; on both occasions he is healthily sceptical, thus:

Whether all those things that we suppose to be bred from corruption and putrefaction, may not be rationally supposed to have their origin as naturally as these Gnats who, 'tis very probable, were first dropt into the Water, in the form of Eggs.

About this time, in Florence, Francisco Redi (1621–97) was performing conclusive experiments to prove that the maggots observed in carrion were derived from eggs laid by blowflies and would not appear if the flies were excluded by gauze. These important experiments were described in his book *Esperienze intorno alla Generazione degl'Inset-ti*[265] (1668), which attracted considerable attention and imitation. Thus, another Italian, Pietro Paulo da San Gallo conducted experiments on the generation of mosquitoes and described them in a letter to Redi dated 1679. In Holland, Stephan Blankaert repeated several of Redi's trials and published them, without acknowledgment, in 1688. Redi did not experiment with lice, but had sound views supported by microscopical observation:

. . . I am more inclined to believe the learned Johann Sperling, that they originate from the eggs of the female, fertilized by coition. And though Aristotle, followed by most moderns, gave out that these eggs or nits as they are called, never produce any sort of animal, he most certainly erred, . . . by the aid of the microscope, they can be seen very well and the full eggs distinguished from those that have hatched.

Redi also published figures of the ordinary and the crab louse (Fig. 6.3) together with various mallophagan and anopluran parasites of animals.[264] His forthright contradiction of Aristotle was all the more remarkable in that he was educated by Jesuits and always tactful in-

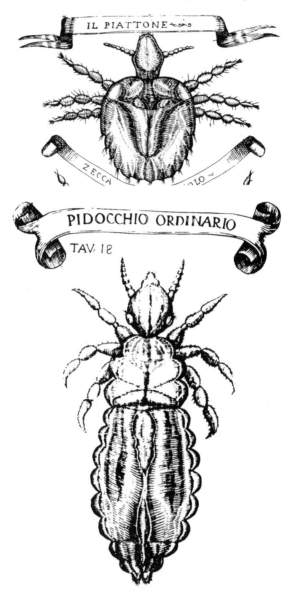

Fig. 6.3. Two illustrations by Francesco Redi,[264] published in 1668. *Above*, a crab louse; *below*, a body louse. By courtesy of The Wellcome Trustees.

dealing with the powerful Roman church. He was a protégé of the Medici Dukes of Tuscany, noted for patronage of art and science, though at the decadent end of their power. Thus, Cosimo II had protected Gallileo, but his successor Ferdinand II yielded him up, as a septuagenarian, to the Inquisition; and Cosimo III, Redi's patron from 1670, was weak, vain and bigoted. To succeed in such a court, Redi needed many talents (he was, indeed, quite a poet) and above all, diplomacy and tolerance. Neither of these qualities were very evident in our next great entomologist, the Dutchman, Jan Swammerdam[278] (1636–80) who was, in fact, invited to the Florentine court, when the Grand Duke visited Holland in 1668 and admired his collection. But Swammerdam rejected the proposal, no doubt wisely; he would have detested the life of a courtier and he held strong nonconformist religious views (verging on the cranky) which would have scarcely accorded with the urbane jesuitical environment.

Swammerdam's father, a prosperous apothecary, intended him for the church; but he, himself, decided that he was unworthy and wished to study medicine instead. He did, in fact, qualify at the University of Leiden in 1663 and displayed much talent in anatomical dissection and the preparation of specimens. In 1672, for example, he dedicated six figures of the human uterus to the Royal Society, and sent them to London, despite the fact that Holland and England were at war at the time! Much of his spare time, however, was spent in natural history studies, initiated perhaps by a substantial collection which his father had amassed. Gradually the fine anatomical work on insects, as well as investigating the life histories of the mayfly and the bee, began to occupy him more and more. He became somewhat estranged from his father (who would have preferred him to cultivate a medical practice no doubt) and somewhat isolated by an egregarious nature and intense religious beliefs; he was often short of money, and plagued by recurrent attacks of malaria. Much of his entomological work was done in the years 1667–9 and his *Historia Insectorum Generis* was published at the end of that period. Much more accumulated in the years up to his death and most of this was published as the *Bybel der Naturae* 57 years later (with German, English and French translations 15 to 20 years after that).

Swammerdam made various contributions to science; but, in regard to the insects with which we are concerned, perhaps his dissections and description of the internal anatomy of the louse is most impressive. Boerhaave, who eventually published the *Bybel der Naturae,* included his biography and some notes on his methods. It seems that he had a brass apparatus constructed, with movable arms to hold his microscope and the object under scrutiny. 'His microscope glasses were of various diameters and focuses ... the best that could be procured in regard to precisions of workmanship and transparency of material.' Anyone who has attempted to dissect a louse (3 mm long!) will realize that lenses alone are not enough; delicate dissecting instruments are essential. To quote Boerhaave again: 'His chief secret seems to have been the making of unbelievably fine scissors, and giving them an extreme sharpness. These he used to cut very fine objects, because they dissected them evenly, whereas knives and lancets, however fine and sharp, are apt to disorder delicate substances ...'

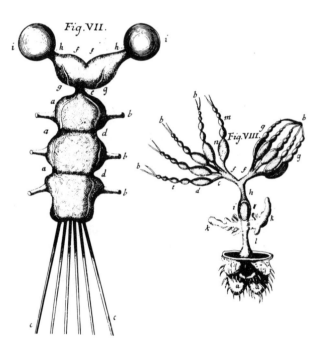

Fig. 6.4. Two examples of Swammerdam's[296] fine dissections of the human louse: ('Fig. VII'), the central nervous system and ('Fig. VIII'), the female reproductive system. (From the *Book of Nature.*) By courtesy of The Wellcome Trustees.

In addition to a clear microscope and delicate instruments, one more thing was necessary: skill in dissecting. There is no doubt that Swammerdam possessed this; yet, at one point, he complains of the difficulty of certain manipulations, in a very human way. Thus, he endeavoured to discern the stylets of the louse in the act of feeding '. . . but I could never accomplish my desire in this particular, though I had then almost wished to have three hands, that I might the better find what I wanted. There are some speculations and researches in anatomy that will not bear writing, since they almost distract the mind.'

In the English translation by Thomas Flloyd,[296] Swammerdam's description of the internal anatomy of the louse occupies over 6½ folio pages (about 6600 words, I estimate). He begins with a discussion of the louse's blood which, he observes, contains globules; and he speculates on an analogy with human blood corpuscles.

Immediately under the skin are certain muscular fibres, which move the annular division of the abdomen . . . Under these muscles the tracheae, or air vessels, come into view; nor could I hitherto discover any vestige of a heart in the upper part of the abdomen, as is usual with other insects . . . this may be owing to its extreme smallness, since it is very difficult to find in the larger insects, as in the Horse-Fly. There is also another impediment, which is the strong and continual agitation of the stomach of the insect, being hardly a moment at rest, from which there arises an unavoidable inconvenience in investigating the heart . . . The ramifications of the trachea, *aspera atria* or windpipe constitute the principal part of this insect; a very considerable number of them are found in the head, breast, belly, legs; nay, and in the antennae or horns.

He discusses these 'pulmonary pipes' in considerable detail, and then:

I shall describe the proboscis or sucker, the throat, stomach, intestines and other adjacent parts . . . The Louse has neither beak, teeth nor any kind of mouth, as doctor Hooke described it, for the entrance into the gullet is absolutely closed; in the place of all these, it has a proboscis or trunk, or, as it may otherwise be called, a pointed and hollow aculeus or sucker, with which it pierces the skin, and sucks the human blood . . .

He also describes quite well the female reproductive system; but for some extraordinary reason, not one of the forty he dissected was a male, so that he was almost inclined

. . . to think that these little animals are Hermaphrodites; and perhaps they really have in each body a penis and an ovary together, as I have found in snails. Whether this be so, is still a secret to me, for though I saw the ovary very distinctly, I could discover no penis . . .

I am tempted to quote Swammerdam much more fully, because it seems to me to be an amazing leap forward, from works of renaissance savants to one which might stand well among those of the late nineteenth century. It is true that other microscopists were active in those progressive years. The renowned anatomist Marcello Malpighi (1628–94) for example, made some excellent dissections of insects, such as the bee and the silkworm; but I do not think he attempted anything as tiny as a louse.

Swammerdam attempted to formulate a new system of classification of the insects, involving four main orders distinguished by the degree of metamorphosis in the life history. The first two orders comprise insects with only partial metamorphosis (without, and with eventual development of wings, respectively); the last two orders comprised winged insects. He naturally assigned the louse, the crab louse and the bed bug to the first order. Later he says:

> . . . I have been told that Doctor Leeuwenhoek observed a Flea at Delft, which, about the end of summer, issued from an egg in the form of a Worm and then shut itself up in a case till the ensuing month of March; but I shall not as yet affirm the certainty of this observation . . . if it did, the insect must belong to the third and by no means to the first order. I shall use the first opportunity of exactly observing this insect, so as to know the certainty of what has been advanced concerning it, as such enquiry cannot be attended with any great trouble.

In this matter of the life history of the louse and flea, the other great Dutch microscopist of the time was somewhat more accurate than Swammerdam. The two were contemporaneous, Leeuwenhoek being born 5 years earlier and living 43 years after Swammerdam. They met at least once, when Swammerdam was a lad, but did not regularly associate, being very different men. Swammerdam, a scholar and qualified physician, lived his rather short and somewhat unhappy life, mainly in Amsterdam in his father's house and latterly as a valetudinarian recluse. Leeuwenhoek (1632–1723) was an untrained amateur, who knew no language but Dutch and lived most of his long life at Delft as a draper.[78] He married twice and had five children, though only one survived infancy; this was his daughter Maria, who kept house for him in his later years. He was not without esteem in his town and was appointed to various minor civic offices. Furthermore, the brilliant results of his microscopical hobby brought him international recognition largely perhaps due to his correspondence with the

Royal Society; so that his house was often visited by curious intellectuals of various nationalities.

Leeuwenhoek was largely self-sufficient; he ground and mounted his own microscope and was, indeed, somewhat secretive about his better models. He left a collection of them after his death, which eventually came (via his surviving daughter) to the Royal Society. The model which has been illustrated in many books, is merely a single lens in a convenient frame; but one must suppose he employed compound microscopes for some work, since he was able to distinguish protozoa, bacteria and spermatozoa.

Leeuwenhoek's microscopical studies of the flea and the louse do not, I think, rival the excellent iconography of Hooke or the amazing dissection details of Swammerdam (Fig. 6.5.). His best achievements in

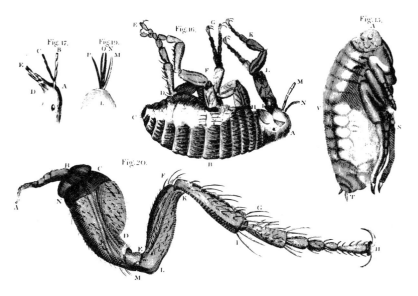

Fig. 6.5. Leeuwenhoek's[180] pictures of the human flea. 'Fig. 15' the pupa; 'Fig. 16' adult flea; 'Fig. 17' and 'Fig. 19', mouthparts; 'Fig. 20', leg. (From *Select Works.*) By courtesy of The Wellcome Trustees.

regard to these insects were his accounts of their life histories, described in an engagingly personal way. Thus, his source of supply of fleas was explained as follows:[180]

During the space of two months and upwards, I put into the hands of my maid-servant four or more glasses with cork stoppers, and directed her to confine in them as many as she could of the largest sort of Fleas, which are always females, with a caution to handle them gently, that none of the veins or vessels in their bodies might be injured . . .

The observations on flea biology were similarly direct and vivid.

In the month of July, I enclosed several Fleas in a glass, that they might lay their eggs; the worms or maggots hatched from their eggs, I nursed with all the care I was able, feeding them every day with flies, which I first killed; these they devoured with great avidity and thereby were speedily increased in size. On the 6th of July, the worm came out of the egg. On the 17th of July, the worm appeared all over white, from which I concluded that it was near dying;. . . but on viewing it with the microscope, I saw that it was employed in spinning around itself a web or covering. The 21st of July, this worm was changed into an aurelia or chrysalis, which was of a transparent white. The 25 of July, this chrysalis assumed somewhat of a red colour . . . The 30th of July in the morning, it was entirely red, and in the evening, the Flea it contained was leaping about in the glass.

The next page is occupied with calculations about the rate of development, the numbers of eggs laid per flea and the duration of the flea's life; in short, the bionomics of the human flea. To sustain the fleas for these experiments, he fed them on his hand, where

. . . one of these sucked the blood with great avidity, standing as it were upon its head, and middle feet with a quivering motion in them.

When he came to investigate similar aspects of body lice, Leeuwenhoek was presented with the disagreeable problem of feeding these rather more disgusting parasites. He says:

. . . I at first proposed to hire some poor child to wear a clean stocking for a week, with two or three female lice in it, and well tied at the garter, to see how many young ones would be produced in that space of time; but I afterwards considered that I could make the experiment with much more certainty in my own person, at the expense only of enduring, in one leg, what most poor people are obliged to suffer in their whole bodies during all their lives. Hereupon I put on one leg, instead of the white under stocking I usually wear, a fine black stocking, choosing that colour, because I considered that the eggs and the young lice thence proceeding, would be more easily distinguished upon it.

After ten days, he finds that one louse has laid about 50 eggs and the other (which has somehow got lost) had laid around 40. The existing louse he dissected and found it still full of eggs.

Having worn the stocking ten days longer, I found in it at least 25 lice of three different sizes, some of which I judged were two days old, and the rest newly come out of the egg, besides others ready to come forth, as I found upon opening one of the remaining eggs. But I was so disgusted at the sight of so many lice, that I threw the stocking containing them into the street; after which I rubbed my leg and foot very hard, in order to kill any Louse that might be on it, and repeating the rubbing four hours afterwards, I put on a clean white understocking.

From his observations, he was thus able to scotch the vulgar saying that 'it (the louse) can be a grandfather, in the space of twenty-four hours.' But nevertheless, his calculations show how large a population can develop, theoretically.

For the pictures illustrating the anatomy of the flea and the louse, he employed a 'limner' who was suitably impressed, exclaiming 'Heavens! What wonders here are in so small a creature!' Unlike Swammerdam, he restricts himself to external morphology, but he relates this to the insect's behaviour. Thus, he notes the function of the 'piercer' and its 'sheath' used for sucking blood; and also the way in which the claw structure is adapted to grasping hairs or fibres of cloth. In one respect he was mistaken; he remarks that the irritation caused by lice is due to the 'sting' carried at the end of the body by male lice. This was, in fact, the penis; and it is surprising that he never observed lice copulating, which would have revealed its purpose.

Leeuwenhoek never wrote a book; his observations and discoveries were recorded in a series of letters, many of them sent to and published by the Royal Society. My quotations are from his 'Select Works' edited by Samuel Hoole and published in 1707. It may be noticed that both Swammerdam and Leeuwenhoek, when wishing to attract wide attention to their work, submitted it to the Royal Society in London. Further evidence of the international coverage of the Royal Society, even in its comparatively early years, is provided by two contributions, describing current Italian work. One of these concerns the life cycle of fleas, as observed by D'Jacinto Cestoni of Leghorn (1637–1718) a pharmacist with great interest in natural history and a friend of Francisco Redi, whom we have already encountered. Extracts of a letter describing Cestoni's observations were published in *Philosophical Transactions* of the Society in 1699,[58] together with the illustration reproduced in Fig. 6.6. The descriptions are brief, but accurate. Cestoni was also involved in an important advance in the study of ectoparasites, with a younger, medical man, Giovanni Cosimo Bonomo (1663–96). Together they

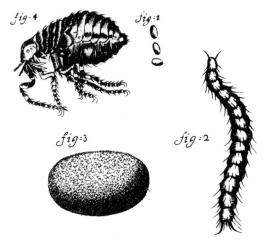

Fig. 6.6. The illustrations to Cestoni's[58] communication to the Transactions of the Royal Society, 1699. By courtesy of The Wellcome Trustees.

investigated the itch mites and I will give an account of their work later (p. 211).

Further evidence of seventeenth century Italian interest in microscopes, and the small creatures revealed by them, is provided by the rather inaccurate pictures of a flea and a louse, published about 1686 (Fig. 6.7). These diagrams, which can almost be considered as the first advertisements for a microscope, have been kindly made available by Dr Mario Coluzzi, having been in his family archives for generations. Not much is known of Lucatelli; he was a student of Pietro da Cortona and worked in Rome, his name being included in the Catalogo Academici Romani of 1690.

Another Italian doctor who made contributions to entomology towards the end of the seventeenth century and at the beginning of the eighteenth, was Antonio Vallisnieri (1661–1730). Educated by the Jesuits like Redi, he came under the influence of Malpighi and decided to study medicine and science. From 1689, he practised medicine in Parma; and, like Aldrovandi a century or so earlier, he founded a botanical garden. From 1700, he was a professor of medicine in Padua; but he still retained his interest in insects and traced the life cycles of several, including the flea, which he illustrated by Cestoni's figures (slightly modified). He also made excursions into the problem of the so

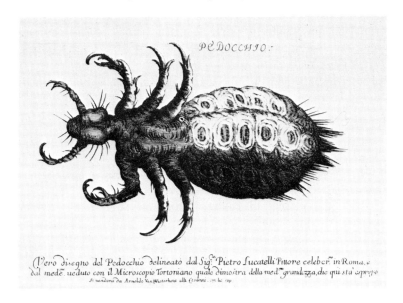

PEDOCCHIO:

(l'ero disegno del Pedocchio delineato dal Sig.re Pietro Lucatelli Pittore celeber.̃ in Roma, e dal mede.̃ ueduto con il Microscopio Tortoniano quale dimostra della med.̃ grandezza, che qui sta' espresso
si uendono da Arnoldo Van Westerhout alla Corona con lie sui

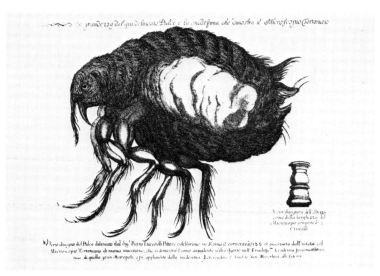

...... in grandezza del qui delineato Pulce e la medesima che dimostra il Microscopio Tortoniano

Vn disegno dell' Istesse
come della larghezza del
Microscopio composto di 3
Cristalli

b) Vero disegno del Pulce delineato dal Sig.r Pietro Lucatelli Pittore celeberrimo in Roma e correntes.̃ 8 & et auertuato dall' istesse col Microscopio Tortoniano di nuova inuenzione, che si dimostra l'ecome antecalente a'dis exposto nell' Eruditi.̃ Accademia Jecometrthone sua di quella gran Metropoli, e fu applaudito dalla medesima Si si rendono si louds a i bei Meschani Ili febori

Fig. 6.7. Pictures of a flea and a louse by Pietro Lucatelli published in Rome about 1686. By courtesy of Dr M. Colluzzi.

called 'lousy disease' (see pp. 103 and 197) and concluded by stoutly maintaining his general principle (*Leggi universali sempre unifromi della gran Madre, spettanti all generazione, propagazione e conservazione de' viventi*)

that every animal is born out of a seed (that in truth is nothing other than its egg) and I must conclude that also these lice which occur in morbid bodies within the flesh are generated thus . . . the Lice of the head are born from Nits (Lendini) and those of the shirt or clothing in the same way, from their eggs.[311]

Vallisnieri was one of the long series of physicians who made contributions to our subject, following those churchmen who first began to dispel the mists of the middle ages. The preponderence of medical men is not surprising, since of the learned professions available at that time, medicine was closest to biology. From the seventeenth century onwards, however, I shall be quoting from some who were not medically qualified. The ingenious and versatile Mr Hooke, for example, could be thought of as an early scientist, though his contributions not mainly in the field of biology. Our next example, however, was pre-eminently a naturalist. John Ray was born in 1627, the son of an Essex blacksmith. He was educated at Braintree Grammar School and later at Cambridge, at the expense of a local squire. He graduated in Greek and the humanities and, ten years later was ordained priest, though he shortly resigned on conscientious grounds. He and his friend Francis Willughby became interested in natural history, Ray being mainly a botanist, Willughby a zoologist. They travelled together extensively in Europe, but Willughby died rather young. Ray subsequently achieved some renown as a botanist and later turned his attention to zoology and entomology. In his posthumously published *Historia Insectorum,*[261] Ray described the feeding habits of fleas, thus:

When they began to feed, they held themselves almost perpendicular and stuck the proboscis into the skin. The irritation did not start at once, but shortly afterwards. When they had filled themselves with blood, they began to eject faeces from the anus; and thus, for many hours if allowed, they suck and eject faeces.

Apart from this, the book contains short notes on the bug, the flea and the louse, but little original on these creatures. On the other hand, Ray's systems of classification both of plants and animals, were distinct advances. Vallisnieri had propounded a system of insect classification based largely on their way of life (living in or on plants, in water, in the earth, or as parasites or predators). Ray's system, while still taking some notice of environment, was primarily based on the type of metamorphosis; thus, he followed and improved upon Swammerdam and led on towards the modern system.

Eighteenth and Nineteenth Centuries

Another early non-medical entomologist was Rene Antoine Ferchalt de Reamur (1683–1757). His many talents made contributions to marine biology, geometry, mineralogy and the digestion of birds; he also devised a thermometer scale; but he is best known for his huge *Memoires pour servir a l'histoire des Insectes*,[262] 1737–48. These deal extensively with butterflies, gall insects, flies, bees and beetles, but not unfortunately with our lowly parasites. I mention him largely to show how the science of entomology was growing in scope. In his introduction he remarks:

... an able botanist must know twelve or thirteen thousand plants! But each one probably has its own special insect, and some, like oak, breed hundreds of different species ...

and he goes on to mention the numerous aquatic species, parasites of vertebrates and the predators. This is a big step from Johannes Sperling who, a century earlier, had been impressed by the great variety of insects, of which he knew some 40 kinds of beetle, 50 grubs, 70 flies and over 100 butterflies!

The greatly widened flora and fauna of the eighteenth century were clearly demanding the improved systems of classification which were to be provided by the famous Swede, Carl von Linné (1707–78). Though he qualified in medicine, practised it for a while and even received a medical professorship, his continual preoccupation was natural history, especially botany. He made contributions to biology and even to applied entomology, but his outstanding work was his systematic classification of plants and animals described in successive editions of his *Systema Naturae*[187] (1735–68). Bodenheimer devotes some pages to ananalysis of the groups of insects classified by Linnaeus in the *Fauna Suecicae*[188] (1746), and the sixth (1748) and tenth (1758) editions of the *Systema Naturae*. In this last-mentioned (in which some 2700 insects are described) the famous binomial system of nomenclature was perfected. The names of several modern orders begin to appear: the coleoptera, the hemiptera, lepidoptera, neuroptera and diptera.

The bed bug is rightly placed in the hemiptera (as Genus VIII *Cimex lectularius*); but the other ectoparasites are relegated to a 'rag-bag' order, the Aptera, which includes spiders, woodlice, scorpions, mites and even crabs, as well as some insects. Linnaeus' system was further improved by his pupil Jean-Cretien Fabricus (1743–1807) who realized the importance of mouthparts in insect classification. He does not, however, say anything of interest on our present subject.

Though we owe the modern scientific classification of animals largely to professionals, the fascinating minutiae of insect biology have often been discovered by amateurs; for example, the ingenious linen draper of Delft. In the early eighteenth century, some entertaining observations on the bed bug were published by another academically unqualified man: a certain John Southall,[291] one of the earliest pest control operators. He describes himself as 'vendor of a nonpareil liquour for destroying Buggs and Nits' and he wrote a treatise on bugs which was of sufficient merit to attract the attention of Sir Hans Sloane, to whom it was dedicated. The frontispiece gives some fairly good figures of bugs of different stages including one from America, which, as might be expected, is larger than the rest. 'But', says Southall, 'Tis useless go give the Gradation of the Species, because when they spawn and breed here, the Young degenerate and are of European size.' Some of his observations were evidently interpreted by a too vivid imagination.

Wild Bugs are watchful and cunning, and though timerous of us, yet in fight with one another, are very fierce; I have often seen some (that I brought up from a day old, always inured to Light and Company) fight as eagerly as Dogs or Cocks and sometimes both have died on the spot. From those brought up tame, I made the greatest discoveries.

On the grounds of a few simple experiments, he makes detailed statements which are often mostly false. Thus, he lists as the food of Bugs:

Blood, dry'd Paste, Size, Deal, Beech, Osier and some other Woods, the sap of which they suck, and on any one of which they will live the Year round. Oak, Walnut, Cedar and Mahogany they will not feed on and starved to death.

On the other hand, Southall was sufficiently observant to refute two common fallacies: that there are no bugs alive in winter; and that only certain people are bitten. Concerning the reproduction in winter, he observes:

. . . in Rooms, in which constant Fires have been kept Day and Night, they have been so brisk and stout as to spawn in the Depth of Winter.' 'But of all the Spawn I ever saw between September and March, not one ever came to Life. This plainly shows that Natural Heat only produces Life in the Spawn and that artificial cannot.

The apparent discrimination of bugs between two people sleeping in the same room he explains by the fact that not all people are marked by the bites. Southall assumes that the marks of bites only appear when the sufferer's blood is out of order.

An amateur of a different type was August Roesel von Rosenhof[269] (1705–59) a fine miniature painter attracted to insects, perhaps, by the challenge of their small size. Fig. 6.8 shows how excellently he depicted the common flea, though it cannot convey his fine colouring. His descriptions of the life stages of various insects were very painstaking and reliable. The value of his publications are somewhat reduced by his ignorance of any language but German, which curtailed his knowledge of earlier authorities.

I suppose one could just describe Charles de Geer (1720–78) as an amateur; certainly he was of independent means and something of a philanthropist. Born of a leading Swedish family, he abandoned a diplomatic career for studies in natural history. It is said that a boyhood present of a silk worm awakened his interest in entomology. He too published under the title *Memoires pour Servir a l'histoire des Insectes*[105] (1752–58), also published in German as *Geschichte der Insekten* (1783). His method, with various species, is to begin with a short list of synonyms with references. The insect is then described, making reference to appropriate copper plate figures (see Fig. 6.9). These appear to be original, but they are only moderately good. His flea, for example, somehow displays four legs; and the broad coxae, containing the jumping muscles, were overlooked (compare Hooke's excellent figure). There is no doubt, however, that De Geer actually examined the insects himself, rather than depending on second-hand information. Though not maintaining lice on his body nor rearing and timing broods of fleas, he does record a few simple experiments. In the very cold winter of 1772, for example, he kept some bed bugs in a powder box in an unheated room to demonstrate their powers of survival. In 1777, he kept fleas in tubes to lay eggs, so that the young stages could be studied. He made careful comparisons of head lice and body lice, deciding eventually that they were distinct species, not merely varieties as Linnaeus had said. The itch mites of which he provides one

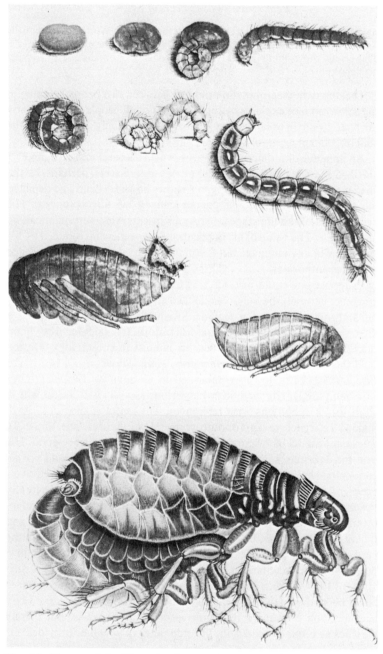

Fig. 6.8. Figures of the life cycle of a flea (probably *Ceratophyllus*) and the adults copulating, by von Rosenhof,[269] 1749. The originals are beautifully coloured.

of the best figures of so difficult a subject, he describes as one of the 'mites, which I had the opportunity of extracting from scabies burrows'.

Errors, of course, appear. He repeats Leeuwenhoek's remarks about the 'sting' at the end of the abdomen of a male louse and says that he himself has seen it. He notes that bugs feed on human blood, but adds that they doubtless feed on other things, since blood is not always available; and he repeats Southall's remarks about their fighting and sucking each other's blood. On the whole, however, in the context of a general text book, the sections on the ectoparasites are reasonably good.

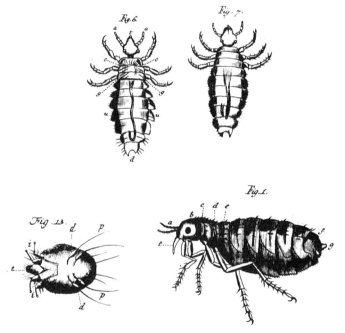

Fig. 6.9. Figures from the History of Insects by De Geer.[105] 'Fig. 6', head louse; 'Fig. 7', body louse; 'Fig. 13', the itch mite; 'Fig. 1', the human flea.

About the same time as De Geer was producing his series of volumes on the History of Insects, a German physician, J. E. Wichmann (1740–1802) published a slim volume on the etiology of scabies (*Aetiologie der Krätze*,[319] 1786). This is mainly of interest in relation to the medical history of ectoparasites and I will refer in a later chapter to his generally enlightened views. At this point, however, we may note his obvious first-hand acquaintance with scabies and the itch mite (including observations of a self-inflicted infection by a colleague); and he produced some quite passable figures of the adult female mite. At the turn of the century, another German medical man, J. H. Joerdans (1764–1813) published a fairly substantial work on the entomology and helminthology of the human body.[158] Joerdans was the son and grandson of doctors and he practised as court physician at Bayreuth. He specialized early in obstetrics but made a few excursions into other fields. I have the impression that, while he must have seen patients infested with ectoparasites, much of the *Entomologie und Helminthologie des Menschlichen Körpers* was based on library learning. Even his figures seem to have been copied from earlier workes, with or without acknowledgement. Thus, his crab louse is copied from Redi, his itch mite from Wiehmann and his fleas from von Rosenhof. He is somewhat inclined to the gruesome and slightly obscene. For example, in his speculations as to infection with crab lice, he not only refers to encounters with prostitutes, but to the sweaty sexual parts of masturbators of both sexes, pressure of fluid in the organs, stinking matter and so forth. Also, he recounts with mediaeval credulity, the tale of a Sicilian who, within a few hours, could fill a box with crab lice from his eyebrows, armpits and groin. 'He had endless irritation, which he increased by scratching. Finally, the male organ became diseased and dropped off; the remainder was amputated, but the patient finally died in great pain.' In a lighter but still indelicate vein, he remarks that fleas are especially liable to attack women, 'whom they can more easily infest as they have more open clothing, giving easier access to their sexual parts.' Presumably, the ladies of those days did not wear drawers; but why the fleas should attack at such a site, the Herr Doctor does not explain.

Errors creep into his accounts of the biology of the ectoparasites. He realizes that head lice feed 'mainly' on the red blood of man ('as seen in their colour and their red-brown excreta'); but of the body louse he says 'only the moisture, perspiration, fat and sticky skin dirt of living men are their food'. Again, he notes that the bed bug feeds on human blood, but he says that its food is not exactly known and it may feed on the sour sap from foul wood; and he repeats Southall's imaginary account of cannibalistic bugs leaving the victims as empty skins. Occasionally, he shows some discernment, as when he correctly identifies the 'sting' of the male louse as its penis. He gives some quantitative data on stages of the flea's early stages, but I am not sure whether these are original or derived from Leeuwenhoek. Characteristically he takes the problem of the 'lousy disease' a stage further, but ends in producing worse confusion (see p. 216); and the same may be said concerning the etiology of scabies.

Early in the nineteenth century, there appeared some general works on entomology by writers whose lives spanned the eighteenth and nineteenth centuries. A substantial article on entomology, by W. E. Leach, appeared in Brewsters *Edinburgh Encyclopaedia* in 1815. This is of particular interest to our subject in that the generic name for crab lice was first established as *Pthirus*. Curiously enough, this seems to have been difficult to spell. Later authors, almost without exception, have added an extra 'h' to give *Phthirus*. This is quite reasonable, since the name is derived from the Greek φθερις. Some writers, however, have added a gratuitous 'i' to form *Phthirius,* for which there is no excuse.

In 1816, J. P. Lamarck (1744–1829) published *Histoire naturelle des Animaux Sans Vertebres;* William Kirby (1759–1850) with W. Spence, published the *Introduction to Entomology* in 1815; P. A. Latreille (1762–1833) appended his *Crustaces, Arachnides et Insectes* to Baron Cuvier's *Regne Animale* in 1829. While all these works have something to say about ectoparasites, they do not introduce much that is original. It is therefore of some interest to quote a non-professional man, a practical man concerned with pest destruction, equivalent to Southall of the previous century. The remarks of Mr Tiffin were recorded by Henry Mayhew who wrote so vividly about the bizarre occupations of the metropolis in *London Life and the London Poor*[203] (1851). This early pest control operator called himself 'Bug Destroyer to Her Majesty' and an account of his contact with royalty is given earlier. He noted that bugs are mostly found in bedsteads, but if unmolested, they get numerous and climb to the tops of rooms and about the corners of the ceilings.

They colonise anywhere they can, tho' they're very high minded and prefer lofty places. They bite all persons the same [?], but the difference of the effect lies in the constitutions of the parties. I've never noticed that a different kind of skin makes any difference in being bitten. Whether the skin is dry or moist, it don't matter. Wherever bugs are, the person sleeping in the bed is sure to be fed on, whether they are marked or not; and as a proof, when nobody has slept in the bed for some time, the bugs become quite flat; and on the contrary, when the bed is always occupied, they are round as a lady bird!

Some people fancy ... the bug smells because it has no vent, but this is fabulous, for they *have* a vent ... They breathe, I believe, through their sides; but I can't answer for that, though it's not through their head. They haven't got a mouth, but they insert the point of a tube, which is quite as fine as a hair, through which they draw up the blood.

It's a dangerous thing for bugs when they are shedding their skins, which they do about four times in the course of a year; when they throw off their hard shell they have a soft coat, so that the least touch will kill them.

I myself have kept bugs for five years and a half without food, and a housekeeper at Lord H—'s informed me that an old bedstead that I was then moving from a store room was taken down 45 years ago, and had not been used since, but the bugs in it were still numerous, though thin as living skeletons.

Tiffin's observations are generally pretty sound, except for his remarks on the survival powers of bugs. (The maximum period survived by starving bugs in modern experiments was 565 days.)

As a complete contrast to these casual field observations are the painstaking laboratory studies of the anatomy of the louse, the crab louse and the bug, made in the middle and later part of the nineteenth century. Outstanding are those of the German doctor L. L. Landois (1837–1902) published between 1864 and 1869. The improved powers of the microscope are evident in his detailed figures; and yet, he was not above some serious errors, as pointed out by the Danish professor, Schjödte,[279] in 1866. Landois followed certain earlier workers (e.g. Burmeister, 1847) in believing that he could see a pair of tiny teeth in the mouthparts of the human louse. These, in fact, do not exist and it is likely that he mistook a pair of thickenings in the head capsule for 'mandibles'. As a result, he concluded that lice could actually gnaw their way under the skin, forming the 'louse abscesses' which figure in the extraordinary story of the 'lousy disease' (see pp. 195–203).

Other microscopic studies in the nineteenth century, include investigations of the louse egg and two Russian workers on the embryology of the head louse and crab louse.

The Modern Era

On reaching the twentieth century, I encounter several difficulties. The profusion of work makes it impossible to write anything like an adequate survey; and this history of scientific interest in ectoparasites is already too long! Only a very brief summary is possible; and if I overlook some important contributors, it must be blamed on my

ignorance or personal preferences. I will group my selection of outstanding contributions under the headings: (i) systematics and
morphology, (ii) bionomics and (iii) biological peculiarities.

Systematics and Morphology

The outstanding work on the bed bug family is the *Monograph of
Cimicidae*[308] (1966) by the late R. L. Usinger *et al.* which summarizes
morphological and ecological work as well as systematics. For the
anoplura, G. F. Ferris' monograph *The Sucking Lice*[88] (1951) very
adequately covers the systematics while G. H. E. Hopkins dealt with
The Host Associations of the Lice of Mammals[145] (1949).

Fleas have attracted many competent workers to study their
morphology and systematics. In the early years of the century, Charles
Rothschild began his enormous collection, incorporating material
supplied by the Indian Plague Commission (1901–6). Rothschild and
his associate Karl Jordan published various papers; and later F. G. A.
M. Smit contributed, all of them being also associated with the flea
collection at Tring Museum. The collection was sent to the British
Museum (Natural History) and remained there many years before a
comprehensive catalogue by G. H. E. Hopkins and Miriam Rothschild,
in five large volumes[147] (1953–71). Other contributors to flea
systematics host associations (and the relation of anatomical
peculiarities to each of these) have been made by G. P. Holland[141] and
especially by Robert Traub.[303]

For a reasonably up-to-date general account of the systematics of
mites, I have turned to the American, G. W. Krantz's *Manual of
Acarology*[174] (1970); but for an authority on parasitic mites I have
relied on the Belgian, A. Fain.[84,85]

Bionomics

Much of the basic biology of bed bugs in this century was elucidated in
German studies. First by A. Hase[118] in 1917 and later in the 1930s,
when he was joined by H. Kemper,[166,167] E. Janisch[155] and E.
Titschack.[302] I 'cut my first teeth' on scientific German, by translating
these long detailed papers. Towards the end of the '30s, C. G. Johnson[160] rounded off the work with an investigation relating laboratory
observations to the natural ecology of the bug in Britain.

Studies of the biology of lice in the twentieth century seem to have been stimulated by the threat of louse-borne diseases in each of the World Wars. At the time of the first war, there were experimental investigations in Germany by A. Hase and Hilda Sikora;[286] while in Britain, G. H. F. Nuttall[165,226] and A. Bacot[18] produced good work. Bacot later studied typhus and contracted the disease, from which he died. Before and during the Second World War, extensive studies were done by my late chief, Patrick Buxton[54] and his collaborators (K. Mellanby, H. S. Leeson and myself).

An extensive investigation of the biology of various fleas was undertaken by A. Bacot,[17] at the suggestion of the Indian Plague Commission, and published in 1914. Though Bacot was a painstaking worker, the knowledge of insect biology and the techniques available then were somewhat elementary, which limits the value of his work. Later studies have been mainly concerned with the tropical rat flea *Xenopsylla cheopis*, which is not only much more easy to colonize in the laboratory than *Pulex irritans*, but is more important as a major plague vector. During the 1930s, useful work on *Xenopsylla* was published by P. A. Buxton and collaborators (H. S. Leeson, K. Mellanby, M. Sharif). Some preliminary biological studies of wild rodent fleas were published during the same era, by the Russian workers, V. E. Tiflov and I. G. Ioff.

So far as the itch mite is concerned, the most advanced studies of its biology and parasitology are those of Kenneth Mellanby[212] and his collaborators in England and of the Dane, Bjorn Heilesen,[123] undertaken during the Second World War.

Biological Peculiarities

One of the most bizarre characteristics of the ectoparasites under consideration is the peculiarity of their sexual organs and mode of copulation. When I was a student, I was taught the 'lock and key theory' of evolutionary divergence; the hypothesis of changes in sex organs being useful to separate emergent species. Certainly the bug, the louse and the flea have evolved most elaborate mechanisms for this vital act. The males of the bed bug family have the curious habit of fertilizing the female by puncturing the body wall, though the insertion is made through a special lateral pit. This was demonstrated by some early work of the Italians A. Berlese and C. Rubaga at the end of the nineteenth century; but a full description was first made by the Englishman, F. W.

Cragg,[70] between 1915 and 1920. The ponderous copulation of lice and the elaborate male penis with its balloon-like inflatable portion was well described by F. G. H. Nuttall[226] in 1917. Another curious facet of the reproduction of lice is that individual families depart widely from the 50:50 sex ratio, as pointed out by P. Buxton.[54] The reason is unknown; but since neither deviation is predominant, they cancel out in large groups of lice.

The complexities of the genital organs of the male flea (which provide useful clues to their identification) have been studied by various workers such as G. P. Holland and R. Traub; and their employment during copulation of the hen flea described by D. A. Humphries.[152] Most male fleas use their stumpy antennae to grasp the female during the act and these are provided with special clasping organs (M. Rothschild and H. E. Hinton).[271]

Other interesting aspects of flea biology are the special arrangements for their phenomenal jumping (p. 16) (Bennet-Clark and Lucey)[24] and, in rabbit fleas, a vaporous stimulation for reproduction derived from their host at the appropriate time of the year (Miriam Rothschild and collaborators).

D. A. Humphries[151] has also contributed various short accounts of the function of combs in fleas, their drinking of water and movements in the cocoon.

7 Medical Aspects of Ectoparasites

VERMIN AS MEDICINE

Primitive medicines are often compounded of highly unpleasant things. Examples could be found from African witch doctors to the South American gauchos, of whom Darwin[73] wrote: 'Many of the remedies used by the people of the country are ludicrously strange, but too unpleasant to be mentioned.'

From the behaviour of young children, I am inclined to believe that there is a psychological reason for this; possibly the masochistic pleasure of overcoming revulsion, or the sadistic experience of imposing this on others! Whether because of the obscure feelings, or because of a long experience of distasteful medicine, many adults used to believe that a good physic should have a foul taste; though this idea is waning with the use of bland modern drugs.

As human intelligence developed, it became imbued with a deep respect for cause and effect; and as it became sophisticated, causes were sought for all kinds of phenomena. Accordingly, some of the ancient philosophers sought a logical justification for unpleasant medicines. In the first century Pliny[246] wrote (*Nat. Hist.* XIX.17),

Certain things, revolting to speak of, are so strongly recommended by our authorities that it would be wrong to pass them by, if it is indeed true that medicines are produced by that famous sympathy and antipathy between things.

Another aspect of the matter emerges from lines in the medical verses of Quintus Serenus Samonicus, about whom little is known except that he was apparently murdered by Caracalla at a banquet in 212 A.D. In one of these verses (5) Serenus[281] points out that lousiness at least prevents laziness!

> *Noxia corporibus quaedam de corpore nostro*
> *Progenuit natura, volens abrumpere somnos*
> *Sensibus et admonitis vigiles incendere curas.*

Cowan's[68] nineteenth century translation runs:

See nature, kindly prudent, ordain
Her gentle stimulants to harmless pain
Lest man, the slave of rest, should waste away
In torpid slumber life's important day.

This utilitarian view of nature was readily transformed into the beneficence of an all-wise Creator with man in mind, which accounted for the popularity of Pliny and some other Roman writers in mediaeval times. Mouffet[219] sums it up, thus:

... that concord and discord, which fills all Physick, by the conduct of nature, hath produced nothing that in some part is not good for man ...

From what I have been saying, it is not surprising to find some insect vermin included in ancient and mediaeval *Materia Medica*. Of the four ectoparasites mainly discussed in this book, the bed bug was most commonly involved, possibly because its peculiar smell made it especially revolting. Lice also feature in some remedies; but fleas and itch mites are notably absent. The latter, of course, are so small and so difficult to obtain in numbers that their omission is hardly surprising. Perhaps this is one explanation for the absence of fleas, as indicated by the following anecdote (from Cowan)[68] which is all that I can find on the subject.

It appears that, in the reign of Elizabeth I, a certain Giles Fletcher (1549–1611) was sent on an embassy to Russia and, in 1591, published a book on his experiences (which was suppressed for many years). One of his stories concerned Ivan Vasilievitch (Ivan III, 1530–1605) who sent a message to Moscow demanding a measure of fleas for a medicine. The baffled citizens replied that this was impossible; and even if they could get them, they could not measure them, because they would leap out. Upon which, he set a mulct upon the city of 700 roubles. This sardonic method of extortion (or taxation, if you prefer a modern term) can scarcely stand as an example of the use of fleas in medicine.

Remedies involving bed bugs can be traced back to Pedanius Dioscorides of Anazarba in Asia Minor, an army surgeon who served in his own country under Nero (54–68 A.D.). The following is an excerpt from his work, translated by John Goodyer (1655):[77]

Cimices of ye bed, (being taken) to the number of seven of them and put in meat with beanes, and swallowed downe before the fitt, doe help such as have ye quartaine ague. And being swallowed downe without beanes, they help such as are bitten by an Aspick. Being smelt unto, they call back such againe as are fallen into a swoune by the strangulation of the Vulua. Being drunk with wine or vinegar, they expel horse leeches. Being beaten small and put into the Urinary Fistula they cure the Dysuria.

Pliny,[240] writing about the same time as Dioscorides also has a number of remedies incorporating bugs (*Nat. Hist.* xix.17).

> The bed bug, a most foul creature and nauseating even to speak of, said to be effective against the bites of serpents, especially asps, as also against all poisons. As proof, they say that hens are not killed by an asp on the day that they have eaten bugs . . . they are applied to bites in the blood of a tortoise . . . fumigation with them makes leeches lose their hold . . . they destroy leeches swallowed by animals, if administered in drink. And yet some actually anoint the eyes with bugs pounded in salt and women's milk, and the ears with bugs in honey and rose oil. Those other virtues attributed to bugs, that they are cures for vomiting quartan and other diseases, although it is prescribed that they should be swallowed in egg, wax or a bean, I hold to be imaginary and not worth repeating. Only as a remedy for lethargy are they employed, with reason, for they overcome the narcotic poison of asps, and are given in doses of seven, with a cyathus of wine and for children in doses of four . . . Furthermore, a couple of bugs attached to the left arm in wool stolen from shepherds, has been said to keep away night fevers and day fevers when attached in a red cloth.

One cannot help wondering why he should be sceptical of bugs as a cure for quartan malaria, if he could accept the nonsense of the last 'remedy'. Elsewhere (xxx.45) Pliny advocates bed bugs for what Mouffet translates as 'moles of the matrix and scabs of the privities.'

Galen[99] repeats the story of bugs being used to expel leeches, but adds that garlic will do just as well (xii.363). According to Mouffet, 'Galen reports that Wall-lice will not only provoke urine, but also drank for nine days space, will stop children's water that goes from them against their will.' (I have not been able to trace this suggestion in Galen's works). Another use of bugs attributed to Galen is to prevent the re-growth of irritating eyelashes which had been plucked out (xiv.349). This curious usage was also mentioned by Pliny (xxx.46); but he preferred the milk of a bitch or the blood of a tick taken from a dog. Q. Serenus Samonicus[281] also mentions tick's blood for this purpose (Cap. 35); but bed bugs figure in some of his other remedies. For quartan fever, for example, he recommends bugs ground up with garlic and drunk in wine (Cap. 49); but for tertian fever, he suggests ground up bugs swallowed in an egg (Cap. 1). Mouffet[219] combines the two versified recommendations as follows:

> Shame not to drink three Wall-lice mixt with wine
> And Garlick bruised together at noon day
> Moreover a bruis'd Wall-louse with an Egge repine
> Not for to take, 'tis loathsome, Yet full good I say.

In a further verse, Serenus advocates bugs to dispel lethargy (Cap 54) which Mouffet translates thus:

> Some men prescribe seven Wall-lice for to drink,
> Mingled with water, and one cup they think
> Is better than with drowsy death to sink.

Mouffet, indeed, collected together a remarkable miscellany of these bug remedies and included some mediaeval ones. For example, '*Valerandus Donures* an Islander, a most learned Apothecary of *Lyons* often said that these drank with water hot, or wine, or both, would wonderfully help those that were troubled with the Stone. Moreover, the later writers wonderfully commend the ashes of them with a fit decoction cast in for a Clyster, to bring forth the Stone.'

Most of these so-called remedies are mere repetitions of fanciful ideas set down by earlier authors without any attempt at verification. But, according to Mouffet, the Swiss physician and naturalist Conrad Gesner actually tried them out. Thus, of the bed bug remedy for the tertian ague, he says:

> Gesner in his writings confirms this experiment, having made trial of it among the common and meaner sort of people in the country.

And again:

> But Gesner, for the Colick prescribes four live Wall-lice to drink in wine in the morning ... and again the next day until they have drank themselves twelve lice ... It helped *Functius* the Governor of *Zurick* at the second taking, and so it did some of his Kindred also, and he was like to have written a commendation of Wall-lice.

Apart from Mouffet, other Renaissance authors repeated many of these ancient remedies; for example, the Italian Matthiolus (1500–77) in his *Commentaries on Dioscorides*,[201] and in England, Humphry Llwyd (1527–68) in his *Treasury of Health*.[159] This was substantially a translation of *Thesaurus Pauperum* of Petrus Hispanus (Pope John XXI) and it includes a remedy for quartan malaria (which is described as 'melancholy putrified'): 'Make a hole in a beane, and put therinto the smale stynkynge wormes that brede in paper or wod call'd Cimices, they take awai the fever'. This book appeared at a time (*c.* 1550) when bed bugs can have been scarcely known in England (see p. 33). We now begin to find remedies involving lice, which were curiously absent in the ancient sources. One of the most persistent was the idea that lice constitute a cure for jaundice. This does not seem to derive from classical sources. Galen, though he devotes pages to jaundice does not mention it: nor does Pliny, who, however, a recommends equally disgusting remedies ('dirt from the ears or teats of a sheep . . . the ash of a dog's head, etc. etc.'). The first reference I can find was that of Aldrovandi[6] who describes it as a remedy among the Belgians; and it is also mentioned by Felix Plater, a Swiss doctor (1535–1614) and repeated in van den Bossche's *Historia Medica*[33] (1689). Mouffet says '. . . experience proves that the jaundies are cured by twelve bruised Lice drank with Wine. *Pennius* gave lice and Butter to beggars and such as live on alms, very often, and so he recovered some that were almost desperate'.

This cure is referred to by Beaumont and Fletcher (1584–1616 and 1579–1625) in the play *Thierry and Theodora*[22] (v. i.) 'Die of the jaundice, yet have the cure about you; lice, large lice, begot of your own dust and the heat of the brick kilns.' And again in William Salmon's *New London Dispensary*[275] of 1676: '*Pediculus* . . . The Louse. They are eaten by Rusticks for the Jaundice and Consumption. Put alive into the Meatus, they provoke Urine.'

Mouffet[219] also says of lice, 'Some for the Dysurie are wont to put into the yard living Lice the greatest they can, to draw forth the urine by their tickling . . .' This is presumably an analogous use to employing bed bugs for this rather bizarre purpose. Then: '*Alexander Benedictus* relates of wig-lice, when clammy humours have hurt the eyes, some cleanse them with Lice put into them, which creeping here and there like *Oculus Christi*, collect the matter; and wrapt up in that they will fall out.'

Alexander Benedictus (d. 1525) was an Italian doctor who practised in Greece and Crete. He introduced the mercurial treatment for syphilis and was noted for wide knowledge of classical medicine. I could not, however, find this particular statement in his works; but the same idea was apparently known to Elizabeth Grey, Countess of Kent, who wrote *Rare Secrets of Physicke*[112] published in 1654: 'For the cure of sore eyes take two or three lice out of your head and put them under the lid.'

During the eighteenth century most of these bizarre remedies began to be discredited among educated people; but some of them lingered on into the nineteenth century, among simple country people. Thus, in the *Lancet* of 1851 there is an account of a French surgeon in Morocco describing a remedy for fever among the natives; if consisted of eating a slice of bread and butter bearing lice. Cowan,[68] too, in 1865, mentioned that in Ohio, the country people sometimes gave bed bugs as a cure for the fever and ague.

A slightly different aspect of insect vermin (notably lice) was the belief that their very presence is beneficial because they remove noxious humours. Galen, indeed, says something to this effect (XVI.290): '*Plurimi igitur ipsos manducantes sudoribus largitur diffluentes corruptum vacuant substantu, ut pediculos quoque vocatus quosdam ejiciant, ex corruptione proprie mascentes, quos ideo, ut autumo phthiras appelant.*' According to Knott,[171] the same idea can be found in Paulus Aegineta (seventh century) and Avicenna (tenth century). Mouffet sums it up thus: 'These filthy creatures . . . are a joy to those that are sick, and sometimes, a cure. For they that have lain long sick of a putrid disease, when lice breed in their heads, they foreshew the recovery of the sick. For it is a sign of the exhaling of it, and flying from the centre to the circumference.' Knott also noted that country people in Ireland believed lice to be beneficial and quoted a French dermatologist who had encountered the same sentiments in peasants in his own country. Even more recently (during the Second World War) when I was attempting to eradicate lice from the heads of refugee children from the Mediterranean, we had definite evidence of at least one mother deliberately reinfesting her child. She thought the lice were beneficial to health and vigour.

REMEDIES FOR ECTOPARASITES

Cleanliness

As we shall soon see, most of the methods of destroying insect vermin, until quite recent times, were ludicrously ineffective. Perhaps the earliest sound remedy was personal hygiene, which I have already discussed (pp. 96 ff). We find occasional good advice in mediaeval and renaissance writers; for example, Bartholomew the Englishman,[19] writing about 1240: 'Against the grieving of lice, oft washing, combing and medicinal cleansing of the hair helpeth . . .' Figure 7.1 reproduces some later woodcuts depicting these operations. Bartholomew also remarked that 'A sluttish kept house breedeth fleas.'

Round about 1500, it was likewise recognized that the best way to discourage body lice was to 'washe oftentymes and to change oftentymes clene linen.'[149] Androvandi,[6] too, noted that bed bugs do not

Fig. 7.1. Hair hygiene to eradicate lice. *Above*, from the *Hortus Sanitatis*[149] (1491); *below*, from the *Historia Medica* of van den Bossche[33] (1639).

breed in beds of which the linen and straw is frequently changed, as in the houses of the rich.

These early writers were right for the wrong reason. They believed that insect vermin were spontaneously generated by dirt and sweat, so that cleanliness would remove their *fons et origo*. There is no doubt that hygienic measures do discourage ectoparasites for various reasons, including destruction of the eggs; so that such measures are still quite useful.

Magic and Religion

If hygiene was the most effective remedy available in ancient times, one may suppose that magical manipulations, incantations and conjurations were the least productive. Furthermore, a magical remedy would be expected to operate immediately and would almost invariably lead to disappointment. For these reasons, no doubt, there are not many records of such attempts. On the other hand, stories of magic have a certain fascination and entertainment value, which may explain the repetition of what is patent nonsense. For example, Pliny retails a contemporary superstition regarding fleas. On hearing the first cuckoo in spring, one should take earth from the ground where one's right foot is standing; this will have the property of driving all fleas from a house. This quaint idea was versified as a mnemonic by one Philes Magorum and Aldrovandi gives both Greek and Latin versions. It was retold in the seventeenth century books, such as Thomas Hill's *Natural and Artificial Conclusions*[132] (1650).

A few magical remedies have found their way into the *Geoponika*[107] a collection of information on agricultural and domestic topics, compiled from Greek and Byzantine sources in the early Middle Ages. As an example of a magic charm, there is a reference to Democritus saying that the feet of a hare, or a stag, hung round the feet of a bed will prevent bed bugs infesting it. The section on fleas includes a ritual in the description of a method of attracting these pests to a dish set in the middle of the room. Round the dish, a line is drawn 'with an iron sword, and it will be better if the sword has actually done execution . . .'

Elsewhere, an incantation (probably derived from the semi-mythical Zoroaster) is suggested: 'If you enter a place where there are fleas, express the usual exclamations of distress and they will not touch you.'

Some writers interpret this as saying 'Ouch! Ouch!' which must be the briefest incantation in record![182]

Somewhat more extensive was the invocation described by the Arabian writer Damiri (1349–91). According to Bodenheimer,[27] he recommended the following way of clearing a room of fleas. A reed smeared with the milk of a she-ass and fat of a goat is set up in the middle of the room and the fleas addressed thus:

I charge you, O Fleas, that you in the name of the Creator of all things, the one and only, the eternal, all-powerful God, that you collect together on this staff. I engage myself by oath and holy vow, not to kill any of you, whether young or old.

This is repeated twenty-five times and the fleas are then presumed to have been charmed on to the reed, which is carefully removed to another place. The room is then washed out and a certain verse of the Koran recited four times.

The beneficial results of sheer holiness are apparent in one of the fourteenth century traveller's tales attributed to Sir John Mandeville[195] (d. 1372). It concerns the monastery on Mount Sinai.

In that Abbey ne entereth not no fly, ne toad, ne newts, ne such foul venemous beasts, ne lice, ne fleas, by the miracle of God and of our Lady. For there were wont to be so many of such filthe, that the monks were in will to leave the place and the abbey and were gone from thence upon the mountain above to eschou that place: and our Lady came to them and bade them turn again, and from thence forwards never entered such filth in that place.

Bodenheimer[27] quotes records of mediaeval and renaissance times describing religious denunciations and objurgations on various agricultural insect pests, though not of domestic insects or ectoparasites. One suspects, however, that these could seldom have had the sensational effect that was witnessed in the celebrated Jackdaw of Rheims!

Trapping

Trapping is not a very efficient method of eradicating ectoparasites but the ingenuity required appeals to some imaginations. For the pests in question, rather different types are required, according to whether the trap is to be worn on the body or used to collect in a room or a bed. It appears that divers flea-traps were worn under the clothes of some elegant ladies in the fifteenth to eighteenth centuries. What was probably the most efficient version, consisted of the skin of a small animal attached to an artificial head, which was worn on a chain round the neck (Fig. 7.2d). One of the heads from such a flea trap, carved from rock crystal and with a jewelled gold mount, was sold at Christies in

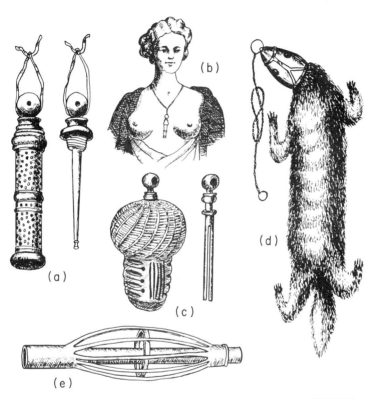

Fig. 7.2. Diverse flea traps. *a*, *b*, as described by Brückmann[37] (1729); *c*, a similar type figured by Madel;[194] *d*, a flea fur with crystal 'head'; *e*, Chinese flea trap described by Eysell.[83]

December 1973 for 8500 guineas. This was said to be 'exceedingly rare, late sixteenth century German'. I was able to examine it and found that it closely resembled the head of the flea fur illustrated by Madel[195] and described as 'Italian, fifteenth to sixteenth century'.

This type of trap might well collect fleas, because of their propensity for burrowing into fur. There are, however, distinct doubts on the efficacy of more sophisticated versions used in the eighteenth century. One of the best known was described by a Dr Brückmann[37] in 1729. It consisted of a small, carved ivory cylinder punctured with small holes, into which a peg smeared with blood was inserted (Fig. 7.2a). This was hung on a cord round the neck and Brückmann's figure illustrates it adorning an attractive bare bosom, which may account for the frequency with which it has been reproduced. Other flea traps of this general type have been preserved in the Heimatmuseum at Butzbach. One is figured by Professor Madel (Fig. 7.2c) and he kindly sent me a wooden model of a similar one, bearing the date 1709. The internal rods of these traps are slotted and it appears that a tuft of wool was inserted and smeared with honey or syrup to trap the intruding fleas.

Flea traps based on the same principle have been used by the Chinese for many years, according to Eysell.[83] As shown in Fig. 7.2e, one form consists of a tube of bamboo, over which a wider, slotted tube has been fixed. The inner tube is coated with bird lime, to which errant fleas might adhere. The Chinese carried these in the loose folds of their sleeves.

In modern times, I was concerned with the development of a personal louse trap, with the late Professor P. A. Buxton.[48] We were working during the Second World War at a time when lousiness was a potential threat, not only to soldiers on active service, but to shelterers during the 'blitz'. The insecticides available, prior to DDT, were only moderately effective and rather unpleasant. We decided to try to concentrate the lice on a treated belt, to be worn under the clothing. To attract the lice to the belt, we had it woven in folds, to take advantage of the strong preference of lice for congregating in seams. It did, in fact, work fairly well on the verminous vagrants we used as 'guinea pigs'; but the introduction of DDT rendered the whole idea obsolete.

When we come to consider flea traps to cope with infested rooms, the problem of a suitable attractant arises. The *Geoponika*[107] recommends digging a pit and filling it with powdered rhodaphne (whatever that is); alternatively, a jar being dug in, with its edge level with the floor, and smeared with bull's fat, is said to attract all the fleas, even those in the wardrobe. This latter method of collecting fleas from a room was repeated in the *Vermin Killer, being a very necessary Family Book*[327] etc. by W. W., published in 1680. The Arab Dimirri, it will be remembered, used a reed smeared with asses' milk and goat fat, but relied on exhorting the fleas to collect on it, as well.

A late mediaeval procedure is mentioned by a citizen of Paris in the

Menagier de Paris[216] written in 1392 as a series of instructions to a young wife.

I have heard tell that if you have at night, one or two trenchers of bread smeared with glue or turpentine and set them about the room with a lighted candle in the midst of each trencher, they [the fleas] will come and be stuck thereto. The other method I have tried and 'tis true: take a rough cloth and spread it about your room and over your bed, and all the fleas that hop thereon will be caught so that you may carry them away with the cloth wheresoever you will.

These suggestions are not without value, particularly the use of a rough cloth in which disturbed fleas would certainly attempt to bury themselves. J. J. Salberg, in a note on *Remedies against Bugs*[274] published by the Swedish Royal Academy in 1745, describes a simple trap for bed bugs used by peasants. It consisted of a plank of spruce wood punched full of little holes; this was placed under the mattress to tempt the bugs with convenient resting places. In the morning, the trap was removed and the bugs killed. Bean leaves have been used in the Balkans for a similar purpose.[29]

A modern bug trap was described a few years ago by a South African.[318] It consisted of a small board (22 × 25 cm) on one side of which a piece of felt was loosely attached. If the trap was hung on the wall of an infested room, bugs gradually accumulated between the felt and the board. The trap was not used to eradicate the bugs, but as an indicator for their presence, especially after control measures had been carried out.

Insecticidal Substances

The number of substances which have been used, or at least recommended for use against insect vermin, would fill very many pages. It would be an immense task to trace the earliest existing records of each suggestion and, even so, these might not be the original ideas. In any case, however, the labour would be unrewarding, for the subject tends to become rather dull for its very inanity. One substance after another was recommended by various writers and repeated uncritically by later authors with more thought of displaying their erudition than of verifying the efficacy of the remedies. Enormous lists were eventually compiled, especially in the renaissance, most of the items being quite worthless.

There is, however, a certain fascination in recounting some of these quaint suggestions and in noting the occasional substances which might conceivably have had some insecticidal action. Among materials of vegetable origin, we find repeated references to *Staphisagria (Delphinium)* which contains alkaloids which have some insecticidal action. Of inorganic materials, mercury figures quite frequently, 'mortified' in various ways; though possibly of some value, its toxicity to man is not negligible. Sulphur was occasionally mentioned; but its most effective uses (in ointment, for scabies, and burnt, as a fumigant) were not mentioned until quite late.

The earliest material remedies for vermin which I can find were listed by the practically minded Romans in the first centuries after Christ. Celsus,[57] for example, accurately describes the infestation of eyelashes by crab lice. His remedies are complicated and drastic (VI. 6): 'In these cases, the bowel should be clystered, the head shaved to the scalp and rubbed for a good while daily, whilst the patient fasts; walking and other exercises should be diligently practised; he should gargle honey, wine, etc., etc.' Finally, he comes to things to kill the lice. 'For this purpose, soda scum, sandarach and black bryony are pounded up together with oil and vinegar ... etc.' Sandarach is a gum obtained from a N.W. African tree; black bryony is a climbing plant in the yam family. I have no idea if they have insecticidal properties. Pliny,[246] also writing in the first century, was considerably more fantastic. For lice, he recommends 'Viper broth, applied to all parts of the body ...' (XXIX. 38) and for nits '... dog's fat, eating serpents cooked like eels or else by bathing their sloughs in drink.' (XXIX. 36).

Galen, indeed, was a Pergamon Greek, but he spent time in Rome, whither he was summoned by Marcus Aurelius. His remedies were somewhat more rational. He recommended resin from ivy against body lice (XII. 30), *Staphisagria* for head lice (XIV. 323) and cedar oil for nits (XII. 17). For crab lice on the eyelashes, he suggested bathing with sea water (XIV. 415). The bland, if not very potent remedy, was repeated by Oribasios (c. 325–400) another Pergamon Greek encyclopaedist and Paulus Aegineta, a seventh century Greek physician. Both these later writers also mention *Staphisagria* for lice.

The *Geoponika*[107] of the early Middle Ages, recommends tar and the juice of the wild cucumber applied to the bed to destroy bugs; or squill cut up and pounded with vinegar. For fleas, an irrigation of *Staphisagria* or pounded leaves of the bay tree having been boiled in sea water is commended.

During the Dark Ages of Western Europe, scientific and medical interests were sustained (as we have noted) by the Arabians. Possibly owing to their interest in chemistry we have inorganic remedies suggested. Thus, according to Mouffet,[219] Avicenna (980–1037) advocates 'Quicksilver with oyl of Roses, and wilde Staves-acre with Arsenick' for destruction of lice. The revival of learning in Europe began fitfully, with such isolated works as that of the Holy Hildegard of Rupertsberg[131] (1098–1179) recommending baked earth, powdered and sprinkled in a bed to eradicate fleas, on the theory that fleas intensely disliked dryness which would make them go away and die. For killing scabies mites in the skin, she recommended applications of extract of mint (*De Plantis* I. 56).

A little later we find remedies in the works of Albertus Magnus.[5] For fleas he recommends an infusion of colecynth or a decoction of brambles; and he says that the odour of the stalks of leaves of oleander repels them. For lice he advocates the juice of lettuce or *Staphisagria* compounded with quicksilver and oil or butter, applied to a belt or to the clothing. Or, alternatively, a powder of quicksilver and lead, thrown on to coals, with the garments held above them for a fumigation. Rather similar ideas occur in writings of a contemporary, Bartholomew, the Englishman, who attributes them to Constantine (1017–1087). '... quicksilver with ashes of willow slayeth them if they be of a hot humour, and so doth lead, burnt with oyle and vinegar; and they be of cold humour, *Staphisagria* and auripigmentum with oyle and vinegar, and so doth sea water'. *Staphisagria* and mercury 'mortified with fastynge spotil' also turn up in the fifteenth century *Book of Quinte Essence.*[256]

By the sixteenth century, when the *Hortus Sanitatis* was compiled, the same remedies, slightly modified, were combined with a few others. In the English book[149] based on this, we have: 'To withdryue the fleen, take alson and ouer rubbe thy bodye every night/ or ellys take thistellys or rewe and sethe that in water and with that water be sprinkell or wasche thy house.'

We have now reached the prolific output of the renaissance period and could cite divers suggestions in a variety of minor works. Even in England I could quote from Thomas Phayre's *Regiment of Life*[241] (1545), Thomas Tusser's *Hundreth Good Points of Husbandrie*[307] (1557) or, somewhat later, from *The Vermin Killer*[327] of the anonymous 'W.W.' (1680). Tusser, for example, put his recommendations into verse, as follows:

> While wormwood hath seed, get a handful or twaine
> To save against March, to make flea to refraine
> Where chamber is sweeped, and wormwood is strown
> No flea for his life dare abide to be known.

The contemporary situation can best be appreciated, however, by an example from one of the great encyclopaedists of the time, such as Aldrovandi[6] who compiled some 1500 words on the control of lice (*Ut fugantur, et occidantur*). A suitable excerpt can be found from the English text of the *Insectorum Theatrum;* this time, on the control of the bed bug.

Against these enemies of our rest in the night, our merciful God hath furnished us with remedies, that we may fetch out of the old or new writers, which will either kill or drive them away. For they are killed with the smoke of Oxe-dung, Horsehair, Swallows, Scolopendra, Brimstone, Vitriol, Arsenic, Verdigrease, Lignum aloes, Bdellium, Fern, Spatula fortida, Birthwort, Clematis, Myrtils, Cummin, Lupins, Knotgrass, Girth, Cypress, as we read in *Aetius, Rhasis, Florentius, Didymus* and *Cardam.* But the best was the curtains drawn about the bed, so to shut in the smoke, that it can have no vent . . . [Omitting another long paragraph] . . . The dregs of boyled Butter cast where there are Wall Lice, will wonderfully kill them for they will feed on it till they burst; May be this is the fat of which *Cardan* speaks in these words; I knew once, saith he, but I have forgotten, a fat, that being smeared on a wooden round circle would so allure Wall Lice to it, like a charm, that one could scarce see the wood for them. Some say that a halfpenny laid under the bed will drive away Wig Lice. Some hang a sheet wet in cold water, and so by actual cold they drive them away. Oyl of itself or with Vitriol, or Bull's gall, or the decoction of a black Chamaelion will drive away Wall Lice. Moreover, all things that are exceeding bitter, or have a stronger smell, will drive them away . . .

After this, it is encouraging to encounter occasional expressions of scepticism in writers of the eighteenth century. There is an interesting essay on 'Experiments to destroy Buggs in Houses' in *Miscellanea vere Utilia*[109] by Boyle Godfrey (2nd edn, 1737). Dr Godfrey said he had met

a learned Professor of Physick and Chymistry, who had been labouring in vain for seven years, who told me he also found the millepede very hard to suffocate, which was his Methods or Attempts to destroy them by: This Gentleman also acquainted me that if Sugar be placed thirty feet from the place where the bugs are, they will run to it; where also a nest of Ants be laid, the Ants will destroy the Buggs; so that if these creatures were not troublesome companions themselves, they would be a Cure against the Buggs. As to all the Pretended

Remedies or Cures against these vermin offered to the Publick daily, I purswade myself they are nothing but Frauds: I let one of those proposers do a bed of mine, when securing a little of his Liquor and examining it, I found it to be Yellow Arsenick and Oyl of Turpentine, which if he cou'd perswade the Bugg to eat wou'd do; but I know he will march over that safely enough. The next famous Remedy that People spent their Money in was Oyl of Turpentine, Camphire and Spirit of Wine, equally to be laughed at.

Godfrey carried out a number of experiments with 'most of the Things found in a Druggists Shop' to endeavour to kill or repel bugs. At last he tried the following, which no Animal Life can subsist under, viz.

Let matches of sticks be made by dipping them in common Sulphur so that they have adhering to them 4 lbs . . . which will do for some rooms . . . which matches must be set upright in earth put into a large Earthen Pot . . . and set on fire in the Room where the Buggs are troublesome, stopping the Chimney by hanging Blankets before it and likewise causing Blankets to be hung against the cracks of the Door . . . This sulphur while burning will give a prodigiously strong Funk and such as will kill all Creatures in the Universe. The Work might be begun in the Morning, and by next Day the Fume will be subsided. It would be proper to make the Fume when the hot weather comes on and they begin to bite; and again in September. All Gold and Silver Laces, Pictures, etc. must be taken from the Room, which with other circumstances renders it troublesome work.

I have quoted this account rather fully because it might serve as sound instructions for a sulphur fumigation, as carried out up to the time of the Second World War, before the advent of modern insecticides rendered the method obsolete.

About the same time, Johan Julius Salberg[274] was experimenting in the same way, in Sweden. He lists many substances of dubious value and selects as the best formula, a mixture of sal amoniac, potash, un-slaked lime and verdigris, distilled with alcohol. On the whole, however, he considered the most effective remedy to be sulphur fumigation.

Sulphur, in the form of an ointment, had been casually noted as an effective cure for scabies by van Helmont,[124] early in the seventeenth century. In the next 200 years, this gradually became more generally accepted (e.g. by Mead, Pringle, Hunter, etc.: see pp. 209 et seq); but even in 1714, Daniel Turner[305] was talking about vipers '. . . boil'd and eaten with their broth.'

It is in the early part of the eighteenth century that we encounter the first English pest control operator in the person of John Southall,[291] whose book on bed bugs I have already mentioned (p. 166). He possessed a secret remedy for destroying bed bugs, which he called

192 Medical Aspects of Ectoparasites

'Nonpareil; and of this, if minded to do it yourself, you may have a bottle for two shillings, sufficient for a common Bed, with plain Directions how to use it effectually.' On the other hand:

If the trouble of doing it yourself be disagreeable to you, you may have it expeditiously done by me or my servants ... on the following terms, viz. To clean a Bed-stead with Moulding-Tester, Wood Head-Cloth, Head-Board and its Furniture, ten shillings and sixpence. Press Beds, Chest of Drawers, Beds and Bed-Steads without Furniture, five shillings each.

Early in his book, he tells how he came to be interested in the subject and about the origin of the 'Nonpareil' liquid. When he was in Jamaica he met an old negress who offered him some liquid to get rid of bed bugs, which caused him much trouble. He found that it worked excellently, driving the bugs out of their harbourages and killing them. He returned to the negress and begged her for the recipe. She took him into the woods and savannah to gather the necessary ingredients. Having returned to London in 1729, he made up some of the liquid and tried it out, comparing its powers with some of the original stuff. Both worked very well, in killing adult bugs; but he found some difficulty in complete eradication since some young stages re-appeared. For this reason, he began his studies on the bug biology. He does not, unfortunately, give any information on the ingredients of his 'Nonpareil' liquid.

About 100 years after John Southall, another professional vermin killer was practising in high society in London. This was Mr Tiffin, whom I have mentioned earlier (p. 171) on account of his observations on bed bug biology, set down later, in Henry Mayhew's *London Life and the London Poor*[203] (1851). Tiffin discusses the problems of eradicating bugs in upper class houses, where it was essential to destroy every last one; but he gives no information on what he used for this purpose. Mayhew's account, however, includes a remedy suggested by a Mr Brande of the Royal Institution. It comprised a mixture of mercuric chloride and ammonium chloride, bound up with yellow wax, turpentine and olive oil. These were among the remedies scorned by Boyle Godfrey 100 years earlier, because, as he pointed out, bugs would not eat such poisons. Nevertheless mercuric chloride was thought worthy of testing against bugs as late as 1912, by Blacklock[26] (who found it worthless).

Mercurial preparations are liable to be toxic to man; indeed that is probably why they were thought to be good insecticides in the first place. Used against bugs, they would not necessarily come into intimate contact with the human body; but until William Buchan[40] warned of their danger, in 1769, they were advocated as applications for scabies. A mercurial ointment was also used until much later, for crab lice. Buxton[54] remarks: 'The use of metallic mercury as a remedy for crab lice goes back to Avicenna; what has already endured for nine centuries will not soon pass out of fashion. Those who persist in following this ancient tradition should be told that mercury ointment is not very effective, and that its lavish use has caused mercury poisoning.'

Since blood-sucking insects will not swallow the poisons we offer them, what is needed is a 'contact poison', or one which kills by mere contact with the insect's cuticle. Only one good insecticide of this type was known during the nineteenth century; this was pyrethrum, prepared from the dried flower heads of a kind of daisy (*Chrysanthemum cinerariifolium*). Its insecticidal powers seem to have been first discovered in Persia and the secret brought to Europe early in the nineteenth century. Later, it was grown extensively in what is now Yugoslavia and the preparation was accordingly known as Dalmation Insect Powder. It formed the basis of what was known in my boyhood as 'Keetings Insect Powder'.

The active ingredients of pyrethrum have the valuable property of causing rapid paralysis to most insects but being almost harmless to man. They (and some synthetic chemicals with similar and even better effect) are still in use today, for purposes where safety and rapid action are paramount; for example, in household fly sprays or in aerosols for aircraft disinsectization. Pyrethrum powder is convenient for treating flea infested pets, especially cats which lick their fur and could be poisoned by more toxic substances.

Pyrethrum, however, has its drawbacks. It is chemically rather unstable and does not give persistent control like modern synthetic insecticides. These were undreamed of in the late nineteenth and early twentieth century. When men began seriously to seek pest control chemicals, they began with familiar poisons (arsenates, fluorides and mercury compounds; and in the vegetable kingdom, nicotine). New compounds were introduced with the growing chemical industries associated with coal tar distillation and the emerging petroleum industry. Many of these (naphthalene, chlorinated cresols, thiocyanates, etc.) were still rather toxic to vertebrates as well as being smelly and unpleasant. Hydrocyanic acid gas and methyl bromide superceded sulphur dioxide; they were more efficient, but more dangerous.

None of these new chemical pesticides provided a satisfactory method of controlling the ectoparasites under discussion. Such progress as was evident was due rather to advances in general hygiene. Thus, lice almost disappeared from Britain in the nineteenth century owing to regular washing facilities becoming more common; as a result, typhus died out completely. Fleas began to disappear in the twentieth century due to improved housekeeping; for example, the wide use of carpet sweepers and vacuum cleaners. When hygiene was lacking, chemicals could not take its place. Thus, despite the resources of the chemical industry, troops in the First World War were persistently lousy. In the slums of great cities, bed bugs remained a persistent and ineradicable aggravation throughout the 1930s.

The situation was radically changed with the introduction of the organochlorine insecticides; DDT, the first and perhaps the greatest, discovered in Switzerland in 1939, followed by gamma BHC, discovered in England in 1942; and later the chlorinated cyclodiene compounds developed in the U.S.A. The special qualities of DDT were its relatively low toxicity to man, which rendered it perfectly safe to handle; its persistence, enabling it to continue killing insects for a long time; and its cheapness, due to its relatively simple manufacture. Its powers were dramatically demonstrated when, in powder form, it rapidly quenched a typhus epidemic in Naples in 1943. After the War, it was used for very many purposes, including an attack on the pernicious bed bug, which it almost exterminated in Europe and N. America.

Insecticides, however convenient and effective, cannot achieve complete eradication of insects, except in limited areas and with exceptional efforts. In the final analysis, we find that human persistence, drive and efficiency are the ultimate answer; and these have been insufficient to get rid of the bed bugs and lice (especially head lice, in children) even in highly civilized countries. In times of national emergency, a special effort is made. During the upheavals of the Second World War, the appalling levels of head lice among city children came to light during the evacuation to country areas. The problems of scabies were revealed in medical examinations of recruits to the forces. Appropriate research in both subjects introduced new and effective medicaments and infestation levels declined. The low levels persisted for years after the War; but in recent times there has been evidence of a resurgence in both scabies and head lice.

Part of the reason for this is a lack of interest in the sordid problem and administrative inefficiency in dealing with it. Some blame may be put on lowered hygienic standards and increased promiscuity among some young people. But a third cause is more serious, because it is less easy to put right. Strains of insects have become resistant (or even immune) to these new convenient and effective insecticides. Bugs in most tropical countries are resistant to both DDT and gamma BHC; this may be spreading to temporate zones. Body lice are resistant to organochlorine insecticides in many parts of the world, thus robbing us of a simple way of stopping typhus. Recently, similar resistance has been found among head lice in Britain and there is little doubt that this situation occurs in other countries, but without being experimentally proved.

The whole subject of resistant strains is a fascinating one and far too complex to discuss here. I will only say that resistance is not due to immunity following non-lethal exposure to the poison (in the way in which disease immunity is produced). It is an innate characteristic in certain individuals in wild populations; and when pesticides are widely and frequently used, mortality of the normal types selects out the resistant forms, which then come to dominate the population. In short, it is a process akin to natural selection ('survival of the fittest') only much more rapid.

The other thing I ought to say is that when a resistant strain has come into existence, it tends to persist, so that the original insecticide cannot be re-introduced. A valuable weapon is lost and the only option is to change to an alternative, of which there are fewer than we would like.

Control of pests by insecticides is not ideal. Improved hygiene maintained by everybody is the best and most reliable answer. Pesticides can help on the way to this ideal; but they must be used intelligently, sparingly and with emphasis on hygiene.

PHTHIRIASIS OR THE LOUSY DISEASE

Of all aspects in the medical history of ectoparasites, the so-called lousy disease is the most puzzling and bizarre. In 1939, a Dutchman, A. C. Oudemans,[229] published what book reviewers would call a 'definitive' account of this curious matter. It runs to some 20 000 words in German, with extracts of original sources in Latin, French, English, Italian and Dutch. I will not, therefore, attempt to rival him, but rather recall the main features of the story and call attention to what I consider a still unsolved problem. Oudemans arranged his very thorough survey chronologically, from about 1200 B.C. to the latter part of the nineteenth century. The first records are remote in time, sparse and often ambiguous. Plutarch,[248] writing perhaps 1200 years later, ascribes the death (in 1190 B.C.) of Akastos, son of Pelias, to this cause. Aristotle,[15] again writing two or three hundred years after the events, cites the philosopher Pherekydos (*c.* 550 B.C.), teacher of Pythagoras. Some mediaeval scholars also include the Cyrenian queen Pheretime (*c.* 575 B.C.); but the earliest citation by Herodotus (484–424 B.C.) suggests rather clearly that gangrene and infestation by fly maggots was involved. The vague classification of maggots and ectoparasites under the heading 'worms' (or vermin) accounts for further confusions of the kind, notably Antiochus Epiphanes (200–163 B.C.), Herod the Great (d. 4 B.C.), Herod Antipas (*c.* 35 B.C.–4 A.D.) and the Roman Galerius Valerius Maximilianus (d. 312 A.D.) all of whom seem to have suffered the fate of Pheretime.

The fantastic stories of the ancients and the equally bizarre tales of their contemporaries were accepted by most medical men of the renaissance. Eventually, however, the physicians, like the early scientists, began to use their own eyes rather than depend on books; and what appear to be sober clinical observations began to be recorded. But the tiny skin parasites ('lice') in dry skin blisters, which are described, bear no relation to any human ectoparasite known today; so that the clarity and consistency of the various descriptions are very puzzling. One point that should be noted is that many of the records are at secondhand; that is, they were set down by people who had not actually seen the cases described.

The earliest detailed records occur in old Chinese medical literature and were discovered by Hoeppli and Ch'Iang[139] (1940). In 1546, about the time when European doctors were writing about slaves sweeping up basketfuls of lice and emptying them in the sea, a Chinese physician called Yi Shuo described two curious cases of a type which was apparently to become prevalent in the West.

A man called Linch'uan, had a tumour on his cheek, which caused unbearable itching ... At last he was told by a physician that the growth was a louse-tumour, which should be opened surgically. His neck was wrapped in oiled paper and the tumour was incised. Numerous small lice were immediately evacuated, followed at the end by two big lice, one white and one black, the size of a bean. The tumour was then empty and no bleeding occurred ... and no cicatrix resulted from the operation.

In the other case,

A student named Li at Fuhang developed a swelling of the size of a cup on his back. It caused no complaint except a peculiar, unbearable itching. One day he was incidentally seen by a doctor, Ch'in Teh-li, who addressed him: 'It is a louse-tumour, I can cure it.' [Some medicine was applied] . . . The tumour ruptured overnight and about one tou (a peck) of lice crawled out from it. The patient said he felt much relieved, but the wound refused to heal up completely. A countless number of lice continued to come out of a small opening over the tumour and the patient finally succumbed.

One of the first circumstantial European accounts was that of Vallisnieri (1733) whom I have already mentioned as an opponent of the theory of spontaneous generation (p. 164). His observations are, however, second hand, being derived from another doctor, Fulvio Gherli, concerning an old man he visited in 1724.[312] The old man's bed was continually covered with lice, despite constant changes of linen. He told the doctor that he had no knowledge of acquiring the lice from anyone, but believed that they were generated in his skin. The doctor

examined as much of his body as possible without causing him embarrassment and found in many places, blisters and swellings. Some were filled with a clear transparent fluid, such as scabies causes; others, however, were dry with a scabies-like scab. I broke them open carefully with a finger nail, so that the fluid would not come out. From the others, I broke off the scab. A fair number of lice were found, some large, some small, lying in a small cave or grotto. On discovering such a wonder, I wrinkled my brows . . .

The next example, from the first half of the eighteenth century, was also recorded at second hand by a Dr M. B. Valentin,[310] who wrote of an unusual case communicated to him by a Jewish doctor of Frankfurt, one Gitman Buxbaum. It concerned a man about 40 who suffered from an intolerable itching which made him tear his flesh with his nails and caused him sleepless nights. He too was covered with a number of small swellings or tubercles. Liberal applications of a decoction recommended by Dr Buxbaum did not cure him and the unhappy man begged him to cut open a tubercle to see what was inside and it seemed to the sufferer that living animals were enclosed.

Led then by curiosity and the hope of relieving the sufferer by a quick surgical procedure, one was opened, and being opened, not a vestige of serum nor a trace of pus was produced, but a large crowd of lice of different sizes and shapes bust forth in countless numbers and the patient, seeing this extraordinary sight, almost fainted. At my advice, the rest of the tubercules were opened and all, like the first, yielded a mass of animaculus and nothing else.

Continuing hence with diaphoretic and cathartic mercurial medicine . . . eventually (with the help of God) the sick man was restored to health.

The next case was also examined in the eighteenth century, though published many years later. W. Heberden in his *Commentaries on the History and Cure of Diseases*[120] (1802) refers to a diary entry of 23 August 1762.

I was this day informed by Sir Edward Wilmot that he had seen a man who was afflicted by the morbus pedicularis. Small tumours were dispersed over the skin, in which there was a very perceptible motion and a violent itching. Upon being opened with a needle, they were found to contain insects in every respect resembling common lice, excepting that they were whiter.

Somewhat analogous to this, is a case in *Ueber lebende Würmer in lebenden Menschen* by Bremser[36] (1819); it was supplied by a colleague, Dr Rust.

In 1808, when I was at the Court of Prince Sangusko . . . I was shown by a Herr Dr Muller . . . a 13-year-old Jewish boy with a swelling on his head, for which many medicaments had been used in vain. On close examination, the greater half of the skull was seen to be covered with a tense dough-like swelling, without undulations, and no trace of inflamation, injury or other abnormality of the skull case. The youth had a slight feverish appearance and complained of an unbearable itching of the swelling . . . by eight days, it had grown to an enormous size. In order to understand the nature of the disease it was decided to open the swelling . . . and there was seen a crowd of small white lice in such quantity as to fill a full Polish quart . . . These were the sole contents of the swelling.

About this time, the German doctor, H. C. Alt, published a dissertation on Phthiriasis[8] (1824). Being convinced that the insects involved in the 'lousy disease' were different from ordinary body lice, he gave them a distinct name, *Pediculus tabescentium*. His description did not, apparently, present any real difference and later authorities (Landois, Murray) have agreed that this is merely a synonym. The name *tabescentium (tabesco,* to erode) however, is of interest in another connection, as we shall see later.

The next case was not recorded at second hand, but was published nearly twenty years after the event. In *Die Krankhaften Veränderungen der Haut*[97] (1840), a Dr C. H. Fuchs recalls a case he saw in 1822 or 1823 in the surgical ward of the Julius hospital in Gottingen

. . . an aged cachetic peasant woman . . . On the skin of the neck and back there were numerous dirty red boils which caused her fearful pain and from which, when she broke them open, there emerged thousands of small lice like mites. The rest of the skin was without boils, though brown, rough and dry with little blotches and dark spots scratched by the patient's nails. She died of marasmus.

Another medical man, Dr Kurtz,[175] described in 1832 the case of a 29 year old woman, a music teacher living in very poor circumstances on inadequate diet.

She sank in strength from day to day and on her skin in many places, rapidly growing boils appeared, slightly inflamed; when these were broken open, countless numbers of lice crept out.

(Kurtz begins his account by dismissing the theory of a certain Dr Stegmann that phthiriasis was induced by excessive sexual drive, especially when it led to abnormal behaviour, such as paederasty. This recalls Joerdans ideas on the origin of crab lice: p. 170).

An Austrian doctor, J. Jeitteles,[156] in 1841 wrote of a patient who came to him a few months earlier.

On closer examination, I found on his rather scaly skin, plentiful small nobs and blisters and lice creeping between them; as when the blisters were burst . . . out of each of them little creatures in masses crept out.

Next we turn to a French doctor Piogey[243] who, in 1853, wrote about a patient exhibited by his lecturer at the Hospital St Louis, Paris, when he was a student. This was a beggar with scabies, syphilis, scurvy and incurable secretion of lice. The skin was covered with pouches, filled with such a large number of lice, that they were obliged to change the sheets every two hours.

Finally we come to the apotheosis of the lousy disease in the writings of the Austrian doctor, Gaulke. These stimulated a lively controversy largely conducted in Viennese medical journals. His findings were sceptically criticized by F. Hebra and also by L. Landois, who called him quite unscientific.

Gaulke had served in the districts on the Russian border and had encountered a good deal of lousiness, which he attributes to low standards of hygiene ('in England, an average 10 lb soap per person is used, in Russia, 1/20th of a pound'). One case described in detail is that of 'Zimmermeister B', examined in 1854, though the account was only published 12 years later.[102]

This man was strong and robust but workshy and a vagabond. He usually returned home lousy and was sent to hospital by the police. Examined by a doctor, he was found to have no special illness, but was anaemic, emaciated and with a yellowish colour. On the skin of the breast and belly were about 100 swellings about the size of a pea or a hazelnut, some open, some covered with thin skin, vivid red, somewhat elevated places, like small abscesses. In the opened places, which reached to the cell tissues, were thousands of lice. The unopened abscesses felt dry as grains, like an atheroma or like the interior of a purse filled with grapeshot or small silver. As soon as these abscesses were opened with a lancet or scissors, one saw a horrible mass of living lice, but not a drop of matter.

As I have said, Gaulke was severely criticized by Hebra[122] (1866), but he remained quite unabashed; 'What I have seen with my own eyes, I will not abandon, despite certain frivolous objections of H'[103] (Hebra). He describes another example, the case of the schoolmaster 'Py', a retired indigent man, semi-blind and obliged to work as a herdsman. He became lousy and did not respond to the usual treatments (soap and alkaline baths). The hospital orderlies reported louse boils, which Gaulke confirmed on examination. He found, as before, many blisters and wounds filled with lice, on his chest. On further examination, more pus-less boils were found, filled with lice, mostly with their anal ends turned outwards

... The nurse 'L' examined these through a lens and found them similar to ordinary body lice from other men, though somewhat more lively. I showed Py to my colleagues, Drs P. G. and B. on 12. xii. 1865. They had never seen such a condition, in spite of some 20 years in the healing profession. It is my experience that a doctor very seldom sees the skin of such a man, unless he has some other illness, as these parasites are usually eliminated by sanitary personnel, who are only too pleased to keep it quiet.

On the other hand, he notes in one of his communications that Hebra had claimed to have seen some 110 cases of lousiness.

L. Landois, who entered into this controversy, although a physician, had devoted much time to a study of the anatomy of lice and bed bugs (see p. 172). One would think that he would be able to dismiss the illusions of the so-called lousy disease. In an early anatomical study of lice, however, he mistook some chitinous bars for mandibles and considered then that lice could, indeed, gnaw a cavity in the skin. This misconception was pointed out by the Scandinavian Schjödte,[279] who dismissed the whole idea of the disease thus: 'Physicians will, I am sure, be not a little pleased to get the ancient monster "Phthiriasis" placed on the retired list, in company with griffins and dragons, the offspring of ignorance.'

Landois, however, adhered to his idea that the gregarious feeding of lice might produce lice pits in the skin.[177] 'Only in one particular of the known covered louse boils will I depart from my earlier conclusion. The lice do not gnaw through the skin but bore into it by their sucking proboscides.'

Subsequent to this time (1866) no one seems to have observed a case which could have given rise to the idea of lice congregating in dry boils. Before discussing Oudeman's solution to the problem, however, I will make a slight digression on the connections between reports of the lousy disease and the idea of spontaneous generation. Up to late renaissance times, the concept of spontaneous generation was so widespread that any reports of lice generated in skin boils would be accepted without question. By the early eighteenth century, however, the English doctor, Daniel Turner, was able to begin his discussion 'Of the Lousy Evil'[305] as follows: 'It is now, in this inquisitive Age, agreed nearly on all Hands, that there is no such thing as equivocal Generation . . .' Nevertheless, he cautions: 'I know that the Incredulous will scarce believe their Eyes, and whether the Overcredulous, especially when prompted to back some Novel Sentiment or Opinion, may not sometimes see what they really do not, I shall leave to others to determine.' In fact, however, the Overcredulous were liable to be backing a mediaeval 'Sentiment or Opinion' rather than a Novel one. For example, in the early years of the nineteenth century (1811) there appeared an extraordinary article by a German medical professor called Wolfart in ΑΣΚΛΗΠΙΕΟΝ a *medicinisch-chirugisches Wochenblatt*.[324] This concerned the 'origin of lice in the hair and the lousy disease in particular.' Wolfart reverts to the ancient theory of the origin of lice in corrupt flesh, apparently due to a disturbance in metabolism. So long, he says, as 'animal vegetation' *(thierische Vegetation)* remains in harmony with other aspects of the totality of an organism, the metabolism remains in balance. But a disturbance results in abnormal growth of parasites. He notes that this can happen in plants too (witness the proliferation of plant lice on unhealthy plants) and that such parasites have a kind of resemblance to their origin (as meal-worms do to meal). Thus, head lice are of the nature and appearance of hair roots. Disturbances of this sort are likely to occur in delicate children, less so in healthy adults but again in the frailty of old age. In short, the whole article is a complete reversion to mediaeval confusion and a sorry contrast to Turner's relatively modern outlook a hundred years earlier.

Some records from the end of the eighteenth and early nineteenth century are inclined to ascribe 'phthiriasis' to infestations of mites. One of these was the Dane, J. Rathke,[259] who published his theories in 1799. The British authors of the *Introduction to Entomology*, William Kirby (Rector of Barham) and William Spence, make a similar suggestion on considering the subject. In their fourth chapter (which they charmingly call '*Letter IV*') they suggest that 'medical men, who were not at the same time entomologists, might easily mistake an Acarus for a Pediculus'. A similar suggestion is made by the German H. Burmeister in his *Handbuch der Entomologie*[42] (1832). It will be noted that the authorities I have quoted have made the suggestion of mites on general grounds rather than in relation to a specific case. Curiously enough, it was Dr Alt[8] (1824), who proposed the species *Pediculus tabescentum*, who describes a case in which, it seems, actual mites are involved. This was a lady called Maria Merzbach, over whose skin tiny animals the 'size of sand grains' crawled rapidly about, causing her considerable irritation. 'I have myself seen these animacules, time and again.' In 1840, another German doctor, G. E. F. Dürr,[80] was still inclined to believe in spontaneous generation of parasites in diseased flesh; but he tended to believe that the lousy disease was due to pus in abscesses turning into mites, rather than lice. Again, A. A. Berthold, in his *Lehrbuch der Zoologie*[25] (1845) assigned the specific name *Sarcoptes tabescentum* to what he described as

Boil mites, body round, brownish, on the back a dark mark 1/3″ long; in cachetic men, in skin boils, from which, when they are broken open, large numbers emerge; it is not certain whether only these or whether yet other mites cause Acariasis.

In 1847, Heusinger (*Recherches de Pathologie Comparee)*[130] observed:

The disease known as *Phthiriasis* consists in the development of animals all over the body, which have not yet been described; these animals are not lice, nor yet Sarcoptes or Acarus, but probably a new genus of mite.

Conclusions. Oudemans, from whose substantial review I have quoted, was inclined to believe in the possibility of 'phthiriasis' being due to mites; and he suggests that the mite responsible was a species of *Harpyrynchus* which he designates *H. tabescentium* (Berthold, 1845). Mites of this genus (e.g. *H. nidulans*) are skin parasites of birds and some species, do, in fact, create little boil-like swellings in which they congregate. Unfortunately for the theory, no one has ever encountered a human case in modern times (Fritsch, 1954).[96] It is true that, at the beginning of Oudeman's article, he mentions an old woman who turned up in 1939 at the University clinic at Utrecht with an extensive parasitic infection vaguely reminiscent of the complaint; but none of the parasites were preserved and the woman, apparently, cured. The parasitic mite theory, is, admittedly, weak; but it is certainly better than involving lice. I have consulted one or two experienced acarologists and dermatologists about the matter (Professor A. Fain; Dr M. Hewitt and colleagues). They have agreed that some *Hyrpyrhynchus* species tend to make pouch-like burrows in animals; and also that mite parasites of animals quite often form temporary infestations on man, causing severe skin reactions. A further possibility is that lice could have been present as a totally separate infestation and so confuse identification. Nevertheless, nothing remotely like the descriptions I have quoted have been actually encountered; so the ancient curse of the lousy disease remains a mystery.

THE SCABIES STORY

Before embarking on the devious history of human attempts to discover the cause of scabies, I ought, perhaps, to outline the nature of the disease.

Scabies is due to the itch mite which spends its entire life cycle in or on the human skin. The only rational way to cure it is by thorough complete skin treatments, which destroy the mites. The largest forms, the adult females, are only a quarter of a millimeter long, but are just visible to the naked eye. They form meandering burrows in the horny layer of the skin, which can be fairly readily seen on careful examination. These are most commonly found on the fingers and wrists, less often on elbows, ankles and a few other sites. The fertilized females lay eggs in their burrows. Young larvae hatch, escape, wander over the body and enter hair follicles. Like the females and other stages, they feed on skin or skin secretions and they eventually moult to nymphs and then to adults. The larvae, nymphs and males appear to invade most parts of the body, though always below the neck. They eventually cause intense irritation, which leads to scratching and causes a rash; then secondary infection by bacteria produces crops of small pustules. The rash occurs in various parts of the trunk and limbs, *but not generally on the hands and wrists where the only easily detectable mites can be found.* This explains the difficulty which doctors encountered in recognizing that the mites were the actual cause of the disease. They simply could not be found in the parts of the body showing the characteristic rash and pustules.

The early history of scabies begins with several distinct threads, representing biology, medicine and folklore; and it was a very long time before these themes were united. The ancients almost certainly recognized scabies and they knew of mites; but not until about 1000 A.D. did doctors notice the presence of mites in the skins of itchy patients, a fact which seems to have been known to peasant women, who dug them out with needles. Another 500 years passed before the mites were recognized as the cause of the complaint, rather than a concommitant phenomenon. Even after this had been reasonably established in the sixteenth century a number of physicians began to question the association, and not until about 1830 was the causation finally demonstrated. The reasons for this slow and fluctuating progress are first the extremely small size of most stages of the mite, which explain the difficulties of scientists prior to the development of efficient magnifying glasses. Secondly, the fact that the most visible stage—the adult female—is usually remote from the areas of the body showing the characteristic rash; and, in particular, these mites are never found in the pustules produced; though one important early account was not sufficiently clear on this point, which misled many later workers.

With a little imagination, we can realize the immense difficulties of the pioneer of medicine. The causes of disease being completely mysterious, it would seem that the obvious remedy was to attempt to placate the gods or demons presumably responsible, with rites, incantations or fetishes. Alternatively, one could appeal to the priests. As I have noted (p. 67) the ancient Jews relied upon priests to distinguish between benign skin infections, such as scabies and the malignant leprosy (Leviticus 13:2).

The ancient Greeks began the process of separating medicine from the web of magic, religion and philosophy; probably Hippocrates (460–357 B.C.) was the first to do this. It must, however, be admitted that many of his theories were somewhat metaphysical. Hippocrates,[135] in fact, considered that all maladies arose from maladjustments of phlegm and bile; though it may be wrong to equate these with modern terms. The O.E.D. interprets the primary usage of phlegm (Gr. φλέγμα) as 'a morbid clammy humour (arising from heat).' Hippocrates grouped scabies with leprosy, prurigo, lichen, favus and alopecia as being produced from phlegm; and he described them more as disabilities than maladies. The Roman Celsus[57] (first century A.D.) contrasted 'scabies' with other pustular diseases. He wrote:

But scabies is harder, the skin is ruddy, from which pimples grow up, some moist, some dry. From some of these, sanies escape, and from these comes a persistent itching ulceration, which in some cases, rapidly spreads. And whilst in some persons it vanishes completely, in others it returns at a certain time of the year. (v. 15)

This description is somewhat vague and it is by no means certain that the disease is that described as 'scabies' today.

Pliny[246] was perhaps more of an encyclopaedist than a physician; and all he has to offer on scabies is a typically bizarre remedy:

For itch in men, the best cure is marrow of the ass or ass's urine with its own mud, butter likewise, which with warm resin also benefits the itch in draught animals. (XXVIII. 75)

Galen[99] like Hippocrates, was born on the eastern coast of the Aegean. He is generally regarded as the acme of classical medicine; but his popularity in the Middle Ages was partly due to the fact that his deterministic views accorded well with Christian doctrine. He held that the human body and its organs were perfectly designed for their functions; convenient evidence of an all-wise Creator. He adhered to the humoral doctrine of disease and lists scabies among other diseases promoted by black bile and melancholy humour (XV. 26 and XVI. 20). He considered that morbid 'ichor' under the skin, turned to solid particles and thus caused various skin eruptions (XV. 22). He notes one significant fact, however, in pointing out that scabies is contagious (VII. 3).

So far, we have been considering the medical views of the Greeks and Romans. For the biological aspect we may cite Aristotle[15] who knew of the existence of mites, which he regarded as a sort of zoological atom, the smallest indivisible organism, an acari. 'A creature is also found in wax [probably a mistranslation for cheese] long laid by, just as in wood, and it is the smallest of animals, and is white in colour, and it is designated the acari or mite.' (*Historia Animalium* v. 32.) Elsewhere (ibid. v. 30) he writes of a kind of louse which can be pricked out of little dry boils in the skin (p. 126). Perhaps these were scabies mites; but, if so, he does not associate them with that disease.

The first to write of the connection of skin-gnawing mites with scabies, was one of the Arabian physicians, who flourished while

Western Europe was still sunk in the Dark Ages. This was Abu Hassan Ahmed ibn al Tabari, physician to the Emir Rukn el Daulah, who flourished about 970 A.D. Parts of his work were translated by Mohammed Rihab (1927) who worked from an Arabian manuscript.[267] Al Tabari followed the humoural doctrines of Hippocrates and Galen and ascribed various forms of 'scabies' to the eating of 'hot, dry' foods; but, for the first time, he described the presence of an animacule in the skin.

This animacule can be removed with the point of a needle. If placed on the nail and exposed to the heat of the sun or fire, it moves. If the animacule is crushed between the fingernails, one hears it crack. This type of scabies is most easily cured . . . by administering laxatives and the killing of the animals.

These words were to be quoted again and again by renaissance writers in the West.

About a century later, came Avenzoa (Ebn-zohr or Ben-Sohr) (1070–1162), quoted by Friedman.[95]

On the external surface of the body, there appears something which the people call 'soab', and it is in the skin. If the skin is removed, there appears from various parts of it, a very small animal, which can hardly be seen. Cleansing of the body may be effected through the use of carthamus seed . . . and the nettle seed destroys this [animal] . . . Nourish him [the patient] well . . . keep away all fresh fruit, especially green or ripe figs . . .

It is noteworthy, in this account, that the itch mites have a common name 'soab', suggesting that afflicted people were well aware of the parasites. As regards remedies. Avenzoa (like al Rihab) combines mite-killing with a dietary regime. Figs, we may note, were often cited as a cause of lice. Joerdens suggested that their numerous pips were reminiscent of tiny lice, mites or their eggs.

At the millenium, the West began to awake from its long intellectual stupor. The German lady Abbess Hildegard (1098–1179) mentions remedies for itch mites. Chapter 52 of her *Physika* concerns the Greater Mint. 'This should be crushed and placed around the place where the *suren* or *snevelzen* are hurting the person with their nibbling, and they will die.' And, the 110th Chapter on Henbane: 'But when the *suren* are in a man, and so sap the strength from his flesh ["dry out" literally] rub the flesh of the affected areas with the juice and the *suren* will die.' Again, we encounter common names for the mites!

Our next authority is Guy de Chauliac (1300–68) who became resident physician to the Papal household at Avignon. In 1363, he completed a large treatise, known as *Chirugia magna*, which became widely used by surgeons for many years. A translation into Middle English was made by an anonymous writer, about the middle of the fifteenth century and a single manuscript of this exists in the Bibliotheque National in Paris.[114] The following are extracts from the transcription by Margaret Ogden (modified by transforming each þ to 'th', for convenience).

Of Scabbe and of ycche Thise forsothe ben bilede infecciones of the skynne, and ycchinge, with scales and with crustes, the wiche ben somtyme altry and ful of quittre and somtyme without it, as Gordonius putteth ... |probably Bernard de Gordon, French physician, d. 1320|. Small pustules bytokene the scabbe, after the same Holy Abbas, begynnynge with ycchinge and afterwards brekynge oute. And thair bytoken a moyste kynde by hete by brennynge, by yychynge and by things that ben put out.

Of lyse and of wormes in the handes and suche others What thing forsothe that a louse is, it is knowen to alle men ... I charge noght forsothe of the maner of gendoringe for it belongeth to phisik. But the thinges helpen to the gendringe of ham whose prepete is to move the mater to the skynne, as ben fyges and lecherie and levynge of clensynge and seldom chaungynge of the clothes.

Hand wormes Ben smale bestes that maken holowe waies and freten betwene the fleisshe and the skynne, namely in the hondes of ydel men.

Little or nothing can be traced on the subject from the next two hundred years; but, with the renaissance came a plethora of publications in which we can find references to the scabies mite. The nineteenth century German dermatologist Hebra collected a number of these. These include Alexander Benedictus (d. 1525) sometime professor at Padua, who published *Omnium a vertice ad calcem morborum*, in 1588; Julius Caesar Scaliger (1484–1558), whom we have already met; he included notes on scabies mites in his *Exercitationes* (1557); Guillaume Rondelet (1507–66) who was born and worked in Montpelier, though he studied in Paris, and was probably the original of Rabelais' Dr Rondibilius. His three-volume *Methodus curendorum omnium morborum humanum* appeared about 1570. The famous French surgeon and physician Ambroise Pare (1509–90) also had remarks to make on the itch mite; and another Italian, Hieronymous Mercurialis (1530–1606) of noble family, born at Forli, published in his *De morbis cutaneis* in 1601.

Most of these references are brief and not original, merely stating that creatures, known in the vulgar tongue as Cyrones, creep about under the skin, causing irritation. Rondeleitius mentions the fact that women extract them with a needle, and Joubertus makes the same remark about common folk. Pare's account is, perhaps, worth quoting (from Hebra):[121]

Les cirons sont petits animaux toussiours, cachez sous le cuir, souz lequel ils se trainent, rampent et le rougent. Ils sont fait d'une matière seiche, laquelle provient du deffaut de viscosité, et diversée et separée comme petits atomes vivants. Les cirons se tuer avec onguent et decoctions faites de choses amere et salées.

These earlier accounts were brought together in the entomological works of Androvandi and Mouffet. The information (true and false) collected by Mouffet[219] and his collaborators is very extensive. He begins

Tho. a Viga falsely reports that the Ancients knew not what Syrones were, for Aristotle calls it *Historia Animalium c.2.* Also they seem to be called Syrones [Greek derivation] because they creep under the skin continually. It is the smallest living creature that is, which useth to breed in old cheese and wax, and also in mans skin . . . in Latine they are called *Pedicelli*, in French, *Cirones,* in Piedmont, *Seiri*, in Gescony, *Brigantes*, in English, *Mites,* in cheese, leves, dry wood and wax; but in men they are called *Wheal-worms.* It dwells so under the skin, that when it makes its mines it will cause a great itching, especially in the hands and other parts affected with them, and held to the fire. If you pull it out with a needle, and lay it on your nail, you shall see it more in the Sun, that helps its motion; crack it with the other nail, and it will crack with a noise, and a watery humour comes forth; it is of a white colour, except the head; if you look nearer, it is blackish, or from black it is something reddish. It is a wonder how so small a creature, that creeps with no feet as it were, can make such long furrowes under the skin. This we must observe by the way, so that these Syrones do not dwell in the pimples themselves, but hard by.

So far, so good; but some equivocal passages begin with an account of Lady Penruddock, who was supposed to have died of a kind of 'lousy disease', but infested with mites (p. 105). Then Mouffet remarks:

It may be some will think it impossible for these Wheale-Worms to breed between the eyes; but we see it is so, and we find it was done so formerly, by an Epistle of *D. Le Ieune,* a Chirugeon to *Jacob Guillimaeus,* his words are these: 'Know, saith he, that in the conjunctive membrane or white of the eye as they commonly call it, some great Wheal-lice by creeping up and down here and there, biting, will make the place itch so much, that a man cannot hold from rubbing.' I do not know what to make of this, nor, indeed, of the following: 'For Syrones in the Teeth: Some call the Worms that breed in mens teeth Syrones, which they affirm have fallen forth like shavings of Lute-strings by the smoke of Henbane seed, received into the mouth. Though I should truly deny that these shavings are Worms, yet Worms breed in rotten teeth Barbers and every man knows.

Despite these rather fantastic passages (recorded at second hand or somewhat dubiously) the description of the itch mite is the best available. From the remark about its colouration, which does not seem to have been copied from other authorities, it seems that Mouffet or one of his collaborators must have actually looked at what they were describing—an unusual habit in early encyclopaedists. Another sound observation was the point that the mites 'do not dwell in the pimples themselves, but hard by'. One would have thought that the way was now clear for a full understanding of the cause, and eventually the cure, of scabies. But medical men were still under the spell of Hippocrates and Galen and the humoural doctrine of disease. To many early scientists, the whole matter was highly theoretical, since they may never have actually encountered the disease; and even for the physicians, it was usually a matter of observing the discomfort of the itch in their patients.

The reactions of one savant who actually contracted the disease, and then suffered the useless torments of early medicine, were quite different. Jean Baptiste van Hellmont (1577–1644) of a good Dutch family, was a highly unusual man. His flashes of genius were often obscured by a torrent of turgid metaphysics. Somewhat a prey to violent enthusiasms he nevertheless devoted much of his life to early chemistry and therapeutics. A follower of Paracelsus, he believed in chemical causes of disease and chemical cures, in strong opposition to the Galenic concept of humours. It is possible that his experiences with scabies set him on this path. The story is told in *Oriatrike, or Physick Refined*, of which an English translation by J. C. Sometime, was published in 1662.[124]

Chapter 51, 'Scabs and Ulcers of the Schools' (referring to schools of medical theory) begins with the words: 'Hitherto I have shown that the causes of Diseases delivered by *Galen* and his followers are erroneous and false.'

He then tells the story of his unfortunate experiences with scabies.

I, being a Young man, about to take leave of a certain Gentlewoman, held her glove and hand for some little while, which laboured with an hidden and dry scab. But I thereupon contracted not a dry but a thin and watery Scab, to wit, onely and that by sober touching; and then I have observed many times that hand towels have brought forth the manginess of scabbed persons . . .

Here, I must pause to remark on his lack of frankness or else his singular misfortune. Modern experiments have demonstrated the very low infectivity of formites or transient bodily contacts in the transmission of scabies. These researches were the inspiration of the following rather scurrilous rhyme.[214]

> Recondite research on Sarcoptes
> Has revealed that infections begin
> When at home with the wife and the children
> Or when you are living in sin;
> Except in the case of the Clergy
> Who accomplish remarkable feats
> And catch scabies and 'crabs'
> On door handles and cabs
> And blankets and lavatory seats.

However young Helmont contracted the itch, he claims to have consulted

two of the more famous Physicians of our City, almost rejoycing that I might now understand in my self whether their Studies answered to their practice. But the Physicians having seen the mattery Scab, presently judged that adust or burnt choler did abound in me, together with a salt phlegm; and so that the faculty of making blood in the liver, was distempered.

Van Helmont, however, was a patient who wanted to discuss the diagnosis and he queried the simultaneous production of a dry and a moist humour ('yellow choler and phlegm'). This seems to have discomposed the doctors who 'being amazed, with their eye-browes bent, they long beheld themselves . . .' and finally the younger one came up with the answer that 'an inflamed Liver did not therefore afford true Phlegm, but abounding salt phlegm, but that the temperature of the salt was hot and dry.'

After further metaphysical speculation, it was decided to bleed and purge the patient. This was done, with great thoroughness (on the fifth day, he records having at least fifteen Stools). Eventually, he declared '. . . my reins were now exhausted, my cheeks had fallen, my voice was hoarse, the whole habit of my body going to ruin, had waxed lean; also it was difficult for me to leave my chamber, and to go, because me knees did scarce support me.'

At least, these drastic experiences made a distinct impression on his opinion of current medical theory and practice. He concluded:

First, That the name of purging was a great deceit. Secondly, That a particular Selection of bringing forth such a humour or any other, was likewise false. Thirdly, Because the birth and existence of humours was also false. Fourthly, That the cause of Scabbedness in respect of burnt choler and salt phlegm was feigned. Fifthly, That the Liver was guiltless in contagions of the skin. Sixthly, That my scab did as yet remain after the purgings, although not with an equal fury. Seventhly, that the fury was not slackened because that one or more imagined humours was expelled ... At length, perhaps after three months, I recovered from my scabbedness by an easie anointing or unguent of Sulphur. Eighth, that the Scab is an affect of the Skin onely ... etc.

Congratulations, van Helmont! Needless to say, he incurred the enmity of orthodox medical men. And the chemical cures for disease turned out to be more difficult to find than he hoped, so that he was unable to save the lives of his wife and two of his children who died young. Since rational medicine proved difficult to demonstrate in reliable cures, the old notions of humoural imbalance persisted for some two centuries and were not infrequently invoked as the cause of scabies. (Van Helmont himself had no conception of the parasitic origin of this complaint. He later ascribed it to an excess of stomach acid which, he thought would produce strangury if it got into the bladder, gout if deposited in the joints and scabies if it reached the skin.)

The discovery and proof of the nature of scabies did not happen suddenly, once and for all. In the seventeenth, eighteenth and early nineteenth centuries, it was demonstrated repeatedly by ingenious doctors; but usually they were disregarded or disbelieved. An important advance was made in the second half of the seventeenth century by the combined efforts of an Italian biologist and a physician. D'Jacinto Cestoni of Leghorn (1637–1718) a pharmacist, with a great interest in natural history, collaborated with Giovanni Cosimo Bonomo (1663–96) a medical man. Together they examined a number of scabetic patients and learnt the trick of extracting the mites, which they examined under what they describe as 'our poor weak microscope'. As I have mentioned earlier (p. 162) Bonomo wrote an account of their discoveries in a letter to Redi, which was published in 1687. Friedman (1936) has published a facsimile of is letter and a translation; but I prefer to quote here from the abridged translation by Dr Richard Mead,[206] published in 1703. This gives the sense well enough, though he changes from the plural to the first person singular; and he omits to say that the person who actually saw the mite was 'Sig. Isaac Colonello (whom we had engaged to draw the figure)'.

Having frequently observed that the Poor Women, when their Children are troubled with the *Itch,* do with the point of a Pin pull out of the Scabby Skin little Bladders of Water, and crack them like Fleas upon their Nails; and that the Scabby Slaves in the *Bagno* at *Leghorn* do often practise this Mutual Kindness upon one another; it came into my Mind to examine what these Bladders might really be.

212 Medical Aspects of Ectoparasites

I quickly found an Itchy person and asking him where he felt the greatest and most acute *Iching,* he pointed to a great many little *Pustules* not yet Scabb'd over, of which pricking one out with a very fine needle, and squeezing from it a little thin Water, I took out a very small white *Globule* scarcely discernable. Observing this with a Microscope, I found it to be a very minute Living Creature, in shape resembling a Tortoise, of whitish colour, a little dark upon the Back, with some thin and long Hairs, of nimble motion with six Feet, a sharp Head, with two little Horns at he end of the Snout; as is represented in Figs. 1 and 3 . . . 'tis hard to discern these Creatures upon the Surface of the Body, nevertheless I have sometimes seen them upon the Joynts of the Fingers in the little Furrows of the *Cuticula,* where by their sharp Head, they first begin to enter, and by their Gnawing and Working in with the Body, they cause a most troublesome Itching, till they are got quite under the *Cuticula.*

With great earnestness I examined whether or no these Animacules laid Eggs and after many enquiries, at last by Good Fortune, while I was drawing a Figure of one of 'em by a Microscope, from the hinder part I saw drop a very small and scarcely visible white Egg, almost transparent and oblong, like to the seed of a Pineapple, as is seen in Fig. 2 and 4. I oftentimes found these Eggs afterwards from which no doubt these creatures are generated, as all others are, that is from a Male and Female, tho I have not yet been able to by any difference of Figure to distinguish the Sex of these Animals.

One would have thought that the observations of Bonomo and Cestoni and their publication in England by Mead would have been sufficient to call medical attention to mites as the cause of scabies. Unfortunately, this was not so. An early English book on skin diseases *De morbis cutaneis* was published in 1714 by Dr Daniel Turner,[305] a London surgeon who published other works on surgery, fevers, etc. Itch mites are considered as a fourth type of louse, in the chapter on the Lousy Evil, as:

Those generated (according to some) under the Cuticle, being found in the Hands and Feet, of a round form like unto the small Eggs of the Butter Flyes, some of them so minute as to escape the Sight, although, by their creeping under the Scarf-Skin, they often stir up a most intolerable Itching . . . Some Authors mention them and treat of them under the names *Acari, Cyrones* and Pedicelli . . .

It is, then, all the more disappointing to find that, in his chapter on the Itch, he makes no mention of skin parasites at all. Instead:

The Cause of this Disease, whether Sympathetic from abroad, or Protopathetic ingender'd in the Blood, is laid in a salt serous Humour, lodged in the Glandules of the Skin, which have been described in our Introduction, raising up the Cuticle into small Pustules or Pimples . . . these for the most part appear between the Fingers (the proper Seat and Pathognomic of the Disease) . . .

As to the origin of this malignant humour, he says:

First of all, the Blood itself being very impure and dissolved, leaves its corruptions and Recrements plentifully in the cutaneous Glands ... Secondly, the Humours gathered in the cutaneous Glands, sometimes be meer Stagnation, become not only itchy but sometimes Corruptive ... Thirdly ... the virulent Infection communicated from without does render if prolifick as to these Diseases ... [so that even those of] as good as Constitution as can be can scarce ever sleep without Harm in the same Bed with an itchy Person, or where such a Person has laid ...

For cures, he admits of 'local Applications or applying Medicines to the Parts. But before these take place, the Blood itself ... is to be freed from Pollution, by bleeding, purging and proper Alternatives ...'

Noting that there is scarce an Old Woman found, without some 'secret for the Itch' he quotes a member from eminent medical authorities, one of whom recommends 'the Flesh of the Viper, dry'd and Powder'd.' Turner himself, however, thinks that this is 'not half so valuable as their Flesh boil'd and eaten with the Broth ...'

We now turn to a later publication by the Richard Mead who published the translation of Bonomo's letter, early in the century, Unlike Daniel Turner, Mead was very sound in his discussion of the Itch, which he says is next to leprosy in foulness but of very different origin.[208] After a description of the signs and symptoms, he wrote:

It may justly be called an animated disease, as owing its origin to small animals. For these are certain insects, so very small as hardly to be seen without the assistance of a microscope, which deposits its eggs in the furrows of the cuticle ... and the young ones, coming to full growth, penetrate into the very cutis, with their sharp heads and gnaw and tear the fibres. This causes an intolerable itching whereby the part is torn and emits a thin humour, which concites into hard scabs. While the little worms constantly burrowing under the cuticle, and laying their eggs in different places, spread the disease.

After discussing possible infection by fomites (linen, wearing apparel, gloves) he adds:

Now what is of greatest moment in this theory, is that the knowledge of the true cause of the disease naturally points out the cure. For neither cathartics, nor sweetness of the blood are of any service here; the whole management consists in external applications to destroy the corroding worms.

For which he recommends warm baths and ointment of sulphur or precipitate mercury.

Admirably modern (except for the mercurial preparation). So too were the 'Observations on the ITCH' by Sir John Pringle (1702–82) in his book on *Diseases of the Army in Camp and Garrison*[252] (1752). According to him, the itch

is of a contageous nature, but the infection is only communicated by the contact of a foul person, his cloaths, bedding, etc. and not by *effuvia*, as in the dysentery and malignant fever. It is confined to the skin and seems best accounted for by *Leeuwenhoek* from certain small insects he discovered in the pustules by the microscope. So that the frequency of the itch is not to be ascribed to the change of air or diet that soldiers are exposed to upon expeditions, but to the infection propagated by a few (who happen to have it at the first setting out) to others in the same ship, tent or barrack.

For a cure, he recommends sulphur ointment.

About this time, a German doctor, J. E. Wichmann (1740–1802), became deeply interested in scabies. He had studied at Göttingen and later in France and England; but he eventually settled in Hanover, where he became a court physician. The studies on scabies were eventually published as a small monograph *Aetiologie der Krätze*[319] (1764) illustrated by copper plates including figures of the scabies mite after Bonomo and the author. The book gives a history of scabies and its mite, mentioning also the flour mite with which it could be confused. The disease is quite well described and also the location of the mites. In conclusion, he states categorically:

I hope I have now thoroughly explained and proved the etiology of scabies, or at least rendered it both plausible and logical that it is a simple skin disease caused by mites . . . The presence of the insects, the cure of the disease by external means, the futility and lack of need of any internal medication, together with the fact that, in this disease one can enjoy a free choice of diet; all of these phenomena are . . . in favour of its parasitic origin.

In a later edition of *Aetiologie der Krätze* (1791) there are references to experimental infections, by putting mites on the skin, carried out by 'G.C.S.U.', a friend of Wichmann, and one Professor Hecker. These describe, in reasonable detail, the early stages of an infestation.

One might well think that the cause and (at least the sulphur ointment) cure of scabies had been well established in the mid-eighteenth century; but I will quote one other supporting medical author. This was a certain John Hunter, who died in 1809; he was not the more celebrated surgeon of that name (1728–93) who articulated the skeleton of the Irish giant (who, indeed, was rather sceptical about the connection of mites and the itch). The John Hunter concerned wrote

Observations on the Diseases of the Army in Jamaica[153] (1788) in which he remarks (p. 291):

While speaking of the diseases produced by insects, it will not be out of place to mention some singularities concerning the itch, a disease which arises from a particular species of insect (*Acarus siro* Linn. Syst. Nat.). It has been doubted whether this disorder really depends upon an insect but I have frequently seen them picked out of the skin, and examined them with a microscope.

He describes the normal progress of the disease and also some very severe neglected cases, which as he says would hardly be recognized as the itch, except that they infected others with ordinary scabies and were themselves cured with sulphur ointment.

It is, as I have suggested, difficult to comprehend how a substantial number of medical men, over the next 60 years, refused to accept the acarian theory of scabies; that is, that mites were the actual cause of the disease. Not only were the form and habits of the mite tolerably well described by naturalists (Mouffet, Leeuwenhoek, Linnaeus, De Geer) but their association with scabies attested by intelligent men (Bonomo, Pringle, Mead, Hunter, Wichmann), One possible reason, which must have impressed clinicians, is the association of scabies, with poverty and its associated poor living conditions, lack of cleanliness and poor diet. Thus, the Scottish doctor, William Buchan (1729–1805) writes in *Domestic Medicine or the Family Physician:*[40]

The itch is a disease of the skin and is generally communicated by infection. It seems originally to proceed from want of cleanliness, bad air, or unwholesome diet; as the inmates of jails, hospitals and such as live upon salted and smoked dried provisions, are seldom free from it.

Rather similar conclusions were reached by the French doctor Anne Charles Lorry (1726–83) in his more specialized book *Tractus de Morbis Cutaneis*[189] (1777). He lists the causes of scabies as: First, cold, damp air; second, poor, indigestible food; third, bad water; fourth, filth generally. Even more categorically than Buchan, he states that all scabies is contagious, being transmitted from man to man by contact, sleeping together or wearing the same clothing. Furthermore, he is aware of the mite theory of causation put forward by Bonomi and supported by Mead; but he cannot believe this for reasons which seem to us vague and irrelevant. Thus, he claims that many cases of febrile disease have been cured by an attack of scabies; furthermore, diseases such as asthma and inflammation have been cured by wearing clothes from a person suffering from scabies. On the other hand, if scabies is 'imprudently driven in,' afflictions of the lungs or other viscera can arise (*Vero scabiei imprudenter retropulse in pulmones visceraque effectus notissimi sunt*). He therefore concludes that scabies is an '*acrimonia sanguis*', residing in the acid and saline serum, which itself contains a principle with a saline taste and a specific penetrating and contagious smell. All this savours of the Galenic humours which so disgusted van Helmont 150 years earlier. And, any way, what of the mites? The answer comes in the thesis of a German doctor, one C. F. Schubert, quoted by Hebra:[121] *Die scabie humani corporis* (1779):

Although I will not deny that worms really exist in the pustules of the itch, yet their presence is no proof that they are the cause. It is quite probable that they are in some way generated by the disease, for we find worms in ulcers and wounds; and yet no one would assert that the worms gave rise to the ulcers. The same thing occurs in achor and eczema of the ears, Tinea capitis, etc.'

In the early years of the nineteenth century, there were several medical authors who discuss the origin of scabies with some care, and, after due consideration, still refuse to believe in the mite theory. One of the most prolix is J. H. Joerdens (1764–1813) whom I have mentioned elsewhere (p. 171) as the author of the very relevant *Entomologie und Helminthologie des Menschlichen Körpers*[158] (1801). This book devotes some 2000 words to a discussion of the nature of scabies, with a fairly reasonable appreciation of Wichmann's contribution. Joerdens points out, however, that several worthy doctors and naturalists have failed to find mites in some cases of scabies. Furthermore, since the mite is known to transfer on clothing (like the germs of the pox, gonorrhoea, etc.) how does it subsist while the clothes are in storage? Furthermore, what about those cases of scabies which arise from excessive external application of medicaments for various diseases, which no doubt mitigate against removal of the itch poison? Joerdens finds it difficult to believe that the mere presence of the mites could cause all the diverse effects of skin diseases. Finally, he considers the fact that scabies can be cured by external medication not critical, since regular applications could well be absorbed and act internally. On the whole, for the cause of scabies, he leans towards malfunctioning of the digestive juices, through unsuitable food (sharp, salty), cold and stagnation of the bowels from an inactive life. True, he feels that the matter should be proved experimentally but he seems fairly sure what the result will be.

One suspects that an actual experiment might not have convinced

Joerdens, with his unshakeable belief in acrid salts and infection by miasmas. We have, indeed, an actual example in the writings of Joseph Adams (1756–1818), especially *Observations on Morbid Poisons, chronic and acute*[2] (1807). While Adams and his friend Mr Banger were visiting Madeira in 1801, an aunt brought him her niece who was suffering from skin mites, called locally '*ouçöes*'. The aunt showed him how to extract these with a needle and both Adams and Mr Banger tried their hand, with indifferent success. They also tried the experiment of infecting themselves with the mites, by putting them on the skin between their fingers. Adams then describes the progress of a scabies infection in himself, the signs and symptoms agreeing well with modern research, except for attacks of a mild fever, which he attributed to the mites (but could, of course, be unrelated).

After all this, however, he concludes that the mites and their skin invasions are not the same thing as the common itch. His reasons, as might be expected, are not very substantial. First, according to vulgar opinion in Madeira, the itch is shameful, whereas the *ouçöes* are commonplace. Next, fever is induced by the *ouçöes*, but does not occur in the itch. Finally, the vesicles caused by the *ouçöes* are regular in form, but irregular in cases of the itch.

Another British authority quite often quoted, is Robert Willan (1737–1812) whose treatise *On Cutaneous Diseases*[321] was completed in 1808. According to Willan, *prurigo mitis* begins when persons neglect themselves and become unclean. If this is allowed to persist, the '*Acarus scabiei* begins to burrow in the furrows of the cuticle and the disease then becomes contagious.' Willan's systematic treatment of skin diseases was continued and expanded by his young friend and colleague, Thomas Bateman (1778–1821), whose *Practical Synopsis of Cutaneous Diseases*[20] appeared in 1813. On *Scabies cachetica* he remarks:

The most ordinary cause of Scabies is contagious, the virus being communicated by the actual contact of those already infected by it, especially where there is much close intercourse . . . Some writers have ascribed the origin of the itch, in all cases, to the presence of a minute insect, breeding and burrowing in the skin; while others have doubted the existence of such an insect.

But, they were both wrong, concluded Bateman, for he had indeed seen such a creature 'taken from the diseased surface', by another practitioner. He himself, however, had never found it and he concluded that there were two kinds of scabies, one with and the other without 'insects'. These anomalies are summarized in Volume 1 of Kirby and Spence's Introduction to Entomology[169] (1815). They are puzzled by the fact that such men as Linné and De Geer, who can surely be relied upon, are convinced that a mite is the cause of scabies. But, they say, 'is it not not equally remarkable that such men as John Hunter, Dr Heberden, Dr Bateman, Dr Adams and Mr Baker should never, in this country, have met with it?'

It will be noticed that the contributors to the scabies story during the latter part of the eighteenth and the first 15 years of the nineteenth century were mainly British or German; but over the next 20 years, a vigorous controversy developed in Paris.[95] It all began at the *Hôpital St Louis* where J. L. Alibert (1768–1837), a life-long advocate of the mite theory, was a professor. Somewhere around 1810, one Jean Chrysanthe Gales, a pharmacist in the hospital, was attempting to graduate as a physician. There is a story that he asked Alibert for a theme for his dissertation and the professor jokingly replied 'Your name is Gales, why not work on *gale* (scabies)?' In any case, this was the subject he chose; and in the course of his research, consulted a number of eminent doctors and entomologists, including the famous Latreille (then about 50 years old). No doubt to the delight of his supervisor, Gales claimed to be able to demonstrate the scabies mite, by extracting them from vesicles of scabetic patients. Furthermore, he appeared to do this at a demonstration attended by Latreille and other eminent authorities whom he had already consulted. Furthermore, he had a cheese mite produced for comparison; and, forthwith, a drawing was made of the alleged scabies mite (by a Dr Meunier of the Natural History Museum). The experts were entirely satisfied and Gales' degree confirmed.

All this seemed very encouraging for those who maintained the acarine theory of the origin of scabies; but then various inconsistencies and even falsities began to appear. To start with, no one was able to repeat this apparently simple operation of extracting mites from scabetic pustules; Gales himself never offered to demonstrate it again in public, though he claimed to have done it over 300 times. According to Friedman.[95]

Alibert had tried, but in vain. So did Rayer Cazenave and Schedel. So did Galeotti at the Hospital of St Eusebe and Chiaruge at the Hospital of Bonifazzil, in Florence. Latreille searched for it among the scabetic patients of the prison of St Denis, but couldn't find it. Mouronval in the clinic of Prof. Lugol working even on holidays and in his recreation time, during the years 1819, 1820 and 1821 . . . finally confirmed complete failure.

Incidentally, suspicion of Gales' performance began to grow, especially when it was noticed that the figure of the so-called scabies mite in his thesis was a clear delineation of an ordinary cheese mite. Eventually, articles ridiculing Gales appeared in the French medical press and the controversy attracted attention outside France; for example, in an article published by the *Lancet* of 6 January 1827. Gales did not bother to defend himself, leaving the defence of the mite theory to the few who remained faithful to it, such as Alibert and Patrix. The sceptics were led by Professor Lugol, who offered a prize of 500 francs to anyone who could unquestionably demonstrate the mite of scabies. The final discrediting of Gales, however, was due to a versatile newcomer, Francois Vincent Raspail (1794–1878), active as a naturalist, a chemist and a somewhat revolutionary politician. He became interested in scabies after hearing Alibert's lecture in 1829 and himself examined many cases. After reading the account of Gales' demonstration he conceived an idea and announced that he would repeat the performance before a critical audience. Arrangements were made for this on 3 September 1829 and, to everyone's surprise, he appeared to demonstrate mites in serum from scabetic vesicles. At the last moment, he showed how he had introduced on his fingers mites from a piece of old cheese in his pocket; and the whole proceedings were later published. It can well be imagined how triumphantly the anti-acarians accepted this demonstration of Gales probable falsity.

But Raspail was far from complacent. He continued studies on scabies of animals, in which he found it easy to isolate mites showing considerable resemblance to the human scabies mite described by De Geer; and in his publications on this subject (1831, 1833) he expressed the opinion that the human scabies mite did exist, though probably more prevalent in warmer climates than that of Paris. He retained the belief, however, that the mite was an exploiter of diseased skin rather than a prime cause.

The final episode in the French scabies controversy was due to a Corsican medical student, S. F. Renucci, of whom, however, little is known. In 1834 Renucci was graduating in medicine and, hearing of the dispute about scabies mites, recalled that in his youth he had seen Corsican women extracting them with a needle (as, no doubt, peasant women had done for ages). He offered to demonstrate the trick and was immediately and repeatedly successful, no doubt to the delight of Alibert who was now 66 years old and must have almost despaired of seeing the mysterious mites. Renucci's demonstration was reported in the *Gazette des Hospitaux* of 16 August 1834. He was at once challenged by Lugol and at a public demonstration 9 days later, removed all doubts about the genuineness of the scabies mite. His success was simply and entirely due to the fact that he knew that the mites were to be found at the end of the meandering burrows and *not* in the vesicles, a fact which had been mentioned by several authors from Mouffet to Wichmann, but which had been overlooked by so many physicians.

Renucci's thesis on the subject includes extensive information on the signs of scabies, the form of the burrows of the mite and the precise method of extracting them. This was published in 1835 (Fig. 7.3). Following his open demonstration, a considerable number of physicians interested in the subject were making investigations, encouraged now by the knowledge of how to obtain mites. Experiments on artificial infestation were done; by Renucci himself and by Albin Gras. An older brother of Renucci, a military surgeon-major, claimed to have cured scabies in a child by carefully extracting every mite. These early experiments were somewhat unsystematic and the proof that mites were the sole cause of the disease had to await the more thorough-going work of Hebra (1844) who strongly maintained 'Without *Sarcoptes,* there can be no scabies. Not all, indeed, were convinced. As late as 1863, Hebra pointed out that there were still some imbued with the idea of scabies as a spontaneous disease; these included such eminent men as Dervegic, Professor of Dermatology at the University of Paris and Drs Cazenave and Schedel, authors of a substantial textbook on diseases of the skin. But, for the most part, the demonstrations of Renucci had set in motion the final series of investigations into the details of the scabies mite and its effects on the human skin. These later works are well summarized in the monograph by the Dane, Bjorn Heilesen[123] (1846). He notes the first substantial advances in morphological studies of the mite by H. Bourguignon (1847 and 1852) of the Veterinary College of Alfort in France. Soon afterwards, further descriptions of the anatomy were made by a German, B. Gudden (1855, 1861), a Frenchman, Charles Robin (1859, 1860) and a Dane, Rudolph Bergh (1860–74). These advances were incorporated in the textbooks of workers themselves engaged in the study, notably Hebra and Koposi (1874) and Megnin (1880).

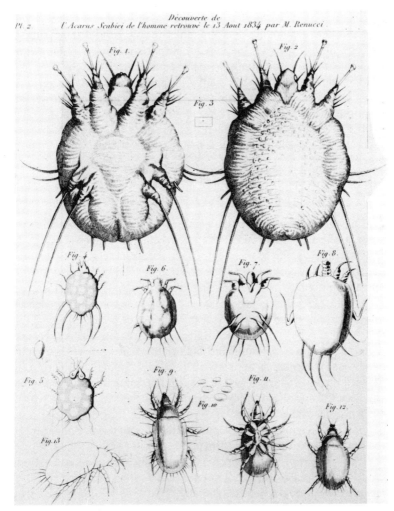

Fig. 7.3. Illustrations of the itch mite and other mites from Renucci's thesis (1834). Courtesy of the British Library.

Soon after the First World War, there were two British contributions, by J. W. Munro (1919) and a detailed study by P. A. Buxton (1921). Attention then lapsed (as, perhaps, also the incidence of the disease) until the Second World War when there were fine contributions from Heilesen, whom I have just mentioned, and Kenneth Mellanby and his collaborators.[212] The former concentrated on anatomical details still awaiting final elucidation and on human reactions. Mellanby and co-workers studied several aspects but their principal contribution was in the epidemiology and transmission experiments. These involved the bedding together of infested young soldiers and conscientious objector volunteers! The main points of the research were summarized in a small monograph (which has just been re-published, virtually unchanged, after 30 years!); but the more human story of the transmission experiments is described in a little non-technical book *Human Guinea-pigs*.[214]

ECTOPARASITES AS DISEASE VECTORS

As I have mentioned earlier, the direct annoyance of blood-sucking insects is of small importance compared with the dangerous diseases transmitted by some of them, in certain circumstances. These include malaria, plague, typhus, yellow fever, sleeping sickness, relapsing fever and various kinds of filariases and encephalomyelitides. Altogether, more than 1500 million people live in parts of the world threatened by one or more of these diseases. It will, however, be noticed that most of the areas affected are tropical; and for that reason, the nature and severity of their toll of human life and health which some of them exact every year, was scarcely appreciated by Europeans until comparatively modern times. Malaria, it is true, was an intermittent lethal scourge in southern Europe, though only as the relatively mild 'ague' in the North. Only plague, carried by fleas, and typhus by lice spread into temperate regions, with terrible historical impact; thus, few of us have not heard of the Black Death, the grisly synonym for plague. Typhus, though equally lethal, tended to be the concomitant of war, starvation and various other camp diseases, so that it did not stand out so sharply.

Curiously enough, both plague and typhus, as well as louse-borne relapsing fever, had largely died out of Europe before their causes were at all understood by the doctors. In other parts of the world, however, they continued to decimate human populations; and it was in these tropical countries that European doctors solved the mysteries of insect-borne diseases, often at the risk of their lives. This is one of the often forgotten benefits of the now much reviled colonialism of the nineteenth century, though not unnaturally, it was the impact of such diseases on expatriate traders, missionaries and administrators which attracted early attention. Malaria gave West Africa its reputation as the White Man's Grave; the construction of the Panama Canal was delayed by yellow fever; African explorers and settlers found their horses and cattle dying of nagana. As a result, the expanding medical and scientific knowledge of the nineteenth century came to grips with problems of a variety of vector-borne diseases. With further advances in the twentieth century, the tropics finally became safe for the white invaders and, in the final years of the colonial era, substantial efforts were made to improve the health of the natives in remote bush and jungle.

By a cross-fertilization of ideas, the discoveries in the aetiology of tropical diseases spread by insects accelerated the understanding of plague and typhus, which had puzzled man for centuries. Before, however, we consider the ways in which insect vectors were incriminated, it may be of interest to look back at this long history of false trails.

A Brief Account of Early Ideas on the Spread of Pestilence

Ineluctable calamities due to chance, or to causes unknown, are difficult for men to accept. Therefore, in all races and at all times, there has been a desire to transfer the onus to a supernatural being who, by analogy with human rulers, is potentially open to appeasement, To my mind, this is the basic reason for all religions; and, in any case, it is relevant to the earliest explanation of epidemics, which were attributed to the wrath of God. The earliest epidemic, which was probably plague, afflicted the Philistines after they had conquered the Israelites (about 1320 B.C.) and it was thought, not only by the Jews, but by the Philistines themselves to be divine vengeance. A similar belief in the wrath of God theory was held by Christians in the Justinian pandemic; and likewise by Moslems who saw in it the will of Allah.

Reactions of ordinary people to this supernatural explanation oscillated between piety and debauchery. The repentance was often excessive and led those with a taste for masochism to the extremities of the fourteenth century Flagellants. But the failure of religion to halt the plague induced in many fatalistic despair and a feeling of 'eat, drink and be merry, for tomorrow we die'. *Britain's Remembrancer*[323] a long poem of some 1000 lines, contains the couplet:

> Some streets had Churches full of people weeping
> Some others Taverns had, rude-revell keeping.

The poem alternates between pious sentiments and observations of the plague in London. It was written by George Wither (1588–1667), a somewhat pious and puritanical man, imprisoned more than once for his trenchant satires.

The consequences of putting the blame on the Almighty were, in any case, less revolting than the actions of those who sought a human scapegoat. Minority groups of foreigners were occasionally the victims, the most obvious choice being the Jews. Stories of plague dissemination by poisoning of wells or by smearing infectious material on doorways, spread like wildfire in the panic conditions of plague epidemics. 'Confessions' were wrung from hostages by horrible torture and mass executions by mobs followed.

These ideas about the dissemination of pestilence have been largely induced and coloured by emotion. It is time to consider the more dispassionate opinions of scholars. In ancient times, these centred round three types of explanation: astrological, miasmatic and contagious. These were not, however, exclusive; and it was quite possible for savants to believe in all three, to different degrees.

Astrology began as an anthropocentric debasement of that most ancient science, astronomy. The irregular movements of the planets, against the apparently fixed pattern of stars, must have been most puzzling to the early observers. The suggestion that they regulated human fate is no more bizarre than divination by examining animal entrails, a practice of Roman augers which has always seemed to me to be peculiarly improbable. The medical associations of astrology were bound up with ancient ideas of the elements earth, air, fire and water, and with their corresponding humours, phlegmatic, choleric, melancholy and sanguine. A sample of such reasoning will suffice, from a work by John Gadbury (1627–1704) published about the time of the 1665 plague. Gadbury was the son of a farmer married to the daughter of a knight, who first disowned her, but later supported John's education at Oxford. He became interested in astrology, published several works and acquired some reputation. Later in life he was accused (apparently falsely) of complicity in Popish plots. The following extract is from his slim book *London's Deliverance Predicted*[98] (1665). 'All Astrologers with good reasons affirm, That all *popular diseases* are irritated by *Mars* and *Saturn* their *Influences:* and, indeed, the skillful in *Sydereal Science* may readily read these dismal effects in their *Natures. Mars* is a Planet *fiery, hot and dry, Choleric;* and therefore Author of all Pestilential Diseases; *Saturn* is a Planet *Earthy, cold and dry* and Author of all *tedious* and *durable infirmities*. And it is observable that *Mars* (though his effects are violent, like his *Nature*) never hurteth so cruelly, or causeth so raging a pestilence, as when in *Configuration of Saturn.*' Apart from the specific effects of planetary conjunctions, the position of *Sirius,* the Dog Star, was considered important, not to mention exceptional events such as eclipses and the appearance of comets.

It seems to me that astrological theories are peculiarly pedantic and must have been devised by introspective scholars remote from everyday events. On the other hand, the origin of disease through foetid miasma, or even more, its dissemination through contagion, seem reasonable suppositions to be made from actual observation. Both hypotheses are extremely ancient. The harmful effects of certain atmospheric conditions were described by Hippocrates, while the dangers of infection are evident in Thucydides' account of the pestilence in Athens (430 B.C.), which may have been typhus. Galen endorsed both theories.

Several factors predisposed men's minds to the idea of disease-bearing miasma. In the first place, several vector-borne diseases are seasonal; notably malaria and to some extent plague and typhus. When an epidemic began over a considerable area, due to favourable climatic conditions for the vector, it would seem as if a noxious miasma had been spread by the wind (the sound wind was commonly blamed). Another suggestive factor is our instinctive aversion to the stink of decay and corruption associated with disease and death; this was even extended to natural stinks such as marsh gases and volcanic fumaroles. Then, as Hirst[136] points out, 'the miasmatic theory made all the greater appeal since it could so readily be linked up with current theological and astrological belief', as Shakespeare did in his three-line summary of the epidemiology of plague:

> Be as a planetary plague, when Jove
> Will o'er some high-viced city hang his poison
> In the sick air. *Timon of Athens* (IV. iii).

The evidence for contagion or infection must have been fairly obvious to any observant physician and by the fifteenth century, the concept was sufficiently sophisticated to distinguish three varieties: by contact, by formites (clothing, etc.) and by infection over short distances. I have said that some early doctors were able to believe in astrological, miasmatic and contagious causes simultaneously; but in later epidemics there was a tendency for individual physicians to adopt one or another theory with great tenacity, so that fierce disputations ensued. These tendencies towards either the miasmatic or the contagious theory had important consequences. Those holding a belief in miasma could do little to counter pestilence other than keep their windows shut when the baneful warm moist southern wind blew; and if the epidemic came, they moved to a more salubrious climate. The contagionists, however, were obsessed with avoiding contact with the disease, so that lepers were banished with bells and plague cases shut up in their houses. In later centuries, the same ideas initiated quarantine laws. On the scientific side, the idea of contagion was decidely more fruitful and set men upon the long and meandering road leading to the discovery of the germs of infectious disease. The story is well told in a paper by Charles and Dorothea Singer,[289] given at a Congress of Medicine in 1914.

The infectious nature of some diseases is much more evident than in others. Galen had recognized it for scabies and phthisis; and the Arabian doctors of the Middle Ages added measles and smallpox. In Renaissance times, venereal pox was an obvious addition and it brought Jerome Frascator (1478–1553) near to the idea of infectious germs. In *De Contagione*[94] (1546) he differentiated between poisons and infections, which could multiply and spread the 'semina' of disease. Other eighteenth century Italian savants' were attracted by this idea of '*contageum vivum*' including Jerome Cardan (see p. 135) and the anatomist Gabriel Fallopio (1523–62); but the name most frequently cited with the origin of the germ theory is probably Athanasius Kircher (1601–80) a German Professor of Hebrew in the Jesuit College in Rome. Among his voluminous writings was *Scrutinium Physico-medicorum Pestis*[170] (1658) in which he suggested that persons infected with plague breathed out minute infective corpuscles; and these he claimed to have seen by microscopic examination of 'putrid' blood from febrile patients.

These early writings are often vague and confused. The concept of *contageum vivum* was found quite compatible with other theories. Thus, Frascator firmly believed in astrological causes of pestilence, while Kircher considered that a foetid miasma was necessary to transform its poisonous corpuscles into living infective agents. Kircher's theories however were developed and promulgated by Christian Lange (1619–60) and attracted considerable attention. Those who believed in the theory of infectious micro-organisms were naturally vague about their nature, since the microscopes available at the time were scarcely capable of revealing bacteria. Such discoveries as had been made with these primitive instruments encouraged analogies with minute worms and insects and in these terms, the infective agents were often described. Others, such as Jean Pestalozzi dealing with plague in Marseilles in 1720, described the infective agent as a volatile saline fermentative principle. This seems to have been the view of Richard Mead (1673–1754) who otherwise held sound views on the subject. Similar opinions persisted into the mid-nineteenth century: in 1866 for example, Benjamin Ward Richardson, lecturing to members of a Congress on Sewage, explained that cholera, typhoid, typhus and diphtheria were due to organic poisons capable of acting as catalytic agents and it was a 'common error to suppose that they multiply from a germ, as offspring from parents.'

It is evident that, from the earliest suggestions of microbes causing disease, there was a long way to go. An important obstacle was the belief in spontaneous generation, according to which germs in diseased tissues were the result of the disease, not its cause. As we have seen, this idea was formerly applied to maggots in putrefying flesh until it was disproved by Redi in the seventeenth century; and it was applied to itch mites in scabetic patients until the early nineteenth century. So far as bacteria and infusoria were concerned, the ancient doctrine had to await demolition at the hands of Pasteur (1822–95).

Proof of the viability of bacteria and revelations of their nature by improved microscopes left a further difficulty in connecting them with specific diseases and proving their causal relationship. The separate identity of some febrile diseases was by no means a simple matter; but despite this difficulty, there were far-reaching advances, which finally gained acceptance. In the closing decade of the century, Robert Koch (1843–1910) had isolated pure cultures of anthrax and with them reproduced the disease in animals. He had also discovered the bacillus responsible for tuberculosis and, with associates, was largely responsible for isolation and description of the germs of cholera, diphtheria and typhoid.

We have now reached the point where it was possible to demonstrate the fact that certain epidemic diseases were transmitted by arthropod vectors. Familiar as we are with this fact, it is difficult to imagine how bizarre and unlikely it appeared to most nineteenth century doctors. As usual, there had been pioneers who had made this suggestion on the basis of more or less tenuous evidence. As early as 1718, the Italian Jerome Lancisi (1654–1730) surveying the malaria infested Pontine marshes, suggested the involvement of mosquitoes (*De Noxiis Paludum Effluviis*).[176] In 1848, an American doctor, Josiah Nott of New Orleans, ascribed both malaria and yellow fever to plagues of mosquitoes. A few years later, Daniel Beauperthay, a French physician living in Venezuela, suggested that the bites of culicine mosquitoes could carry the poison of yellow fever from stagnant waters to the human blood stream. This theory was further improved by Carlos Finlay in Cuba in 1881 to include man-to-man infection by mosquitoes. About the same time, Patrick Manson, as Medical Officer to the maritime customs in China, actually demonstrated the development of a nematode parasitic to man, in the bodies of mosquitoes.

The numerous advances at the turn of the century are better known. In 1895 David Bruce in Zululand associated nagana with tsetse flies and eight years later, in Uganda, extended this to sleeping sickness. Ronald Ross, studying malaria in Bangladore, demonstrated oocysts on a mosquito's stomach (1897); this work being further elucidated by the Italians, Grassi, Bignami and Celli. The U.S. Yellow Fever Commission in Cuba finally incriminated *Aedes aegypti* as the vector, by human experiments in Cuba (1900). These crowding discoveries form the background to the discoveries of the transmission of plague, typhus and relapsing fever.

The Elucidation of Plague

Not only do the effects of bubonic plague cause widespread alarm, but the signs and symptoms are generally so distinct that few epidemic diseases have been more easily recognized. On occasions, indeed, there has been official reluctance to diagnose it, by generals who feared for the morale of their troops, or by civilian authorities concerned for the disastrous effects of quarantine on commerce. Nevertheless, it usually managed to claim recognition eventually and it is one of the most extensively documented diseases in history. At the time of the Black Death and for some years afterwards, hundreds of 'plague tractates' appeared (281 have been reproduced in *Archiv der Gesichte der Medizin* between 1910 and 1925). British contributions to the subject were common during or immediately after the great epidemic of 1665–6; 15 are listed in Payne's edition of the contemporary treatise of Boghurst. Among modern treatises, Hirst's *The Conquest of Plague*[136] is particularly interesting in tracing the evolution of understanding of the disease. The earliest speculations on the cause of plague were, as might be expected, perfused with magic and metaphysics. Later, more rational men favoured theories of contagion or noxious miasma, though neither explanation was consistently convincing, even with the limited evidence available in ancient times. This is because of the protean nature of plague, comprising the flea-borne bubonic disease, in which the patient is virtually non-infectious, with intervening episodes of the highly infectious pneumonic form.

Some polemical views on these opposing theories appeared in the plague tractates; but in these early times, the controversy was obscured by alternative mystical explanations. It was not really until the sixteenth century that the divergence into miasmatists and contagionists began to crystallize. In discussions of the Venetian epidemic of 1576, for example, Mercurialis (1530–1606) upheld the miasmatic theory in *De Pestilentia*[217] (Chap. 17), blaming pestilential vapours from the lagoon. Whereas Massaria (1510–98) supported the idea of infection in *De Peste*,[200] being perhaps impressed by the progress of epidemics carried from one Italian city to another.

In Britain, the controversy was stimulated by a series of epidemics in London, in 1563, 1593, 1603, 1625, 1636 and culminating in the Great Plague of 1665. Among contemporary historians, William Boghurst (1631–85) supported the miasmatic doctrine in his book *Loimographia*,[30] being particularly impressed by its simultaneous appearance in different parts of the City.

The Plague or Pestilence is a most subtle, peculiar, insinuating, venomous, deleterious Exhalation, arising from the Faeces of the Earth extracted into the Aire by the heat of the sun, and difflated from place to place by the winds, and most tymes gradually but some tymes immediately aggressing apt bodies.

The opposing doctrine of contagion was advanced by Nathaniel Hodges (1629–88) in *Loimographia* and by Daniel Defoe (1661–1731) whose *Journal of the Plague Year* is well known. These contagionist views were later endorsed by Richard Mead (1673–1754), whom we have met elsewhere (p. 211, 213), who, like Hodges, was convinced that plague was imported from the Levant on cotton or silk goods. Mead was very interested in the plague epidemic in Marseilles (1720–1) which he discussed in his *Short Discourses concerning Pestilential Contagion*.[207] This epidemic, which was initiated by the arrival of an infected ship from Syria, was investigated by an official Commission, appointed by the Regent of France, on the advice of his personal physician, M. Chirac. The Commission, drawn from members of the Medical School at Montpellier, diplomatically endorsed the miasmatic views of M. Chirac; but they 'hedged their bets' by recommending various anticontagion measures. The epidemic was exceptionally severe locally, causing nearly 40 000 deaths among Marseilles' 90 000 inhabitants; and it caused great interest and alarm in other western countries, which further stimulated controversy. The problem was made the subject of a competition and the winner, Jerone-Jean Pestalozzi (1674–1742) was a confirmed contagionist.

In the nineteenth century, plague gradually receded from Europe, though British doctors were much concerned with an epidemic in Malta in 1813–14. When, however, plague broke out in Egypt in 1834–5, numerous European doctors in the country became involved and the old controversy was revived again. Actual experiments were conducted to try to settle the matter. Criminals were inoculated with blood from plague patients or made to sleep in their underclothes. Two French doctors, Clot-Bey and Bulard themselves took part in such risky trials; but the matter was not clearly settled.

The final medical onslaught on the problem of plague took place at the end of the nineteenth century in connection with the last great plague pandemic. This began, as we.have seen (p. 58) in Central China and, after reaching the coast, was carried to many parts of the world. During the coastal epidemics, notably in Hong Kong, in Bombay and in Sydney, plague was studied yet again by doctors of many nationalities. This time, the advances in medicine and science during the nineteenth century resulted in the final elucidation of plague; but not without some complications and difficulties.

To begin with, the plague bacillus was discovered in 1894 in Hong Kong, almost simultaneously by two men. Shibasaburo Kitasoto (1856–1931) a Japanese bacteriologist, who had studied in Berlin under Koch, had been sent to Hong Kong with Professor Aoyama, a pathologist, by the Japanese Government. Alexandre Yersin (1863–1943) from French-speaking Switzerland, had worked at the Paris Pasteur Institute and was despatched to Hong Kong by the French Colonial Minister. Both discovered the pathogen in tissues of plague victims, soon after their arrival. Kitasoto's announcement was slightly earlier, while Yersin's description of the bacilli rather clearer and it is his name which is associated with the modern name of the bacillus, *Yersinia pestis.*

Incrimination of the pathogen responsible for plague still left unsolved the problem of its dissemination. Not only was there no knowledge of the involvement of fleas, but the importance of rats was yet to be demonstrated. One would have thought that the epizootic deaths of rats which precedes an outbreak of bubonic plague would have been noticed repeatedly, even in ancient times. It is, indeed, possible to find occasional references to this association; for example, Strabo (*c.* 50 B.C.) who noted that in Spain, rats frequently give rise to pestilence and Avicenna who described how rats and other subterranean animals left their burrows and wandered about in a drunken manner, during plagues. But most of these early mentions, and certainly those recorded at the time of the Black Death or during the Great Plague of London, deal indiscriminately with all kinds of animal life (rats and mice, dogs and cats, birds, etc.) which were said to be afflicted and, by some authors, accused of being disseminators of plague.

Perhaps the first scientific records of rat deaths, at the beginning of plague epidemics, date from the latter part of the nineteenth century and come from colonial and missionary doctors in the Far East. But these observations had no significance for the considerable array of doctors engaged in the study at this time. In Bombay alone, at the turn of the century, there were British, Austrian, German, Egyptian and Russian Plague Commissions as well as a local Bombay Plague Committee. An immense amount of work was undertaken by members of these bodies, at considerable risk. (One of the Austrian doctors contracted pneumonic plague in subsequent research and died.) But, as Hirst remarks, the scientific achievements of organized bodies often are less successful than the efforts of certain gifted individuals. I have already mentioned Kitasoto and Yersin; the latter, with E. Roux,[328] first formulated the theory of rat involvement in 1897. 'La peste, qui est d'abord une maladie des rats, devient bientôt une maladie de l'homme. Il n'est qu'une bonne mesure prophylactique contre la peste serait la destruction des rats.' About the same time, Professor M. Ogata[228] of the Hygiene Institute in Tokyo, proved the existence of plague bacilli in fleas taken from an infected rat; and he suggested that fleas as well as other insects might act as vectors.

These hypotheses were developed in a brilliant way by P. L. Simond (1858–1947). Born in Southern France, he qualified in medicine at

Bordeaux and worked at the Pasteur Institute at Marseilles, under Metchnikoff. From 1896 to 1900 he served in the Far East and witnessed the initiation of the plague pandemic in Yunan before working in Bombay. From close reasoning on the epidemiology of plague, he showed that the chain of infection was much more likely by rats than by human case contacts, or by aerial or water transmission.[287] The clue which led him to the idea of flea transmission was the observation of local lesions caused, in some cases, by the bites of infected fleas. These small greyish spots ('phlyctenes') always heralded the appearance of a bubonic swelling in the affected region of the body. Simond carried out experiments in transmission from rat to rat by fleas. with moderate success; and he became convinced that this was the mode of transmission. It explained why handling a rat newly dead of the plague was so dangerous; but, within an hour or two, when the parasites had left the cold cadaver, it could be handled with impunity.

If these early theories could have been followed up, the conquest of plague would have been expedited. Much time and money could have been saved by abandoning the spraying of plague houses with strong disinfectant, a practice which tended to spread the disease by driving out infected rats to adjoining premises. Unfortunately many doctors, and especially the commission members, were unimpressed. The eminent parasitologist, G. H. F. Nuttall[224] (1862–1931) himself carried out some transmission experiments with entirely negative results, largely because he expected fleas to become infectious immediately after feeding on infected rats. In Hong Kong, W. Hunter also concluded that infected fleas were unimportant in plague transmission. In Marseilles, however, J. G. Gautier and A. Raybaud managed to confirm Simond's work; and the British Commission for Investigating Plague in India, working from 1905 to 1908, gradually reached the same conclusion.

Various important details remained to be discovered; for example, the relative efficiency of different species of flea as vectors; this largely depended on the readiness with which they would bite man as well as rats. The European rat flea, *Nosopsyllus fasciatus*, is reluctant to bite man; whereas the human flea, *P. irritans*, will not easily feed on rats. Neither are normally vectors of plague, which is most commonly spread by fleas of the genus *Xenopsylla*, especially *X. cheopsis*. Another factor is the liability of different fleas to suffer gut blocking by plague bacilli (see p. 56). This phenomenon and its importance was largely demonstrated in 1914 by the remarkable British naturalist, Arthur

William Bacot (1866–1922), whom I shall mention again in relation to typhus.

With the elucidation of the bacillus, rodent and flea involved, the main problem of plague control was solved. Further advances in the next half century were pharmaceutical and entomological rather than medical. In the 1930s, sulphonamides were introduced and these were found effective for treatment, though now largely displaced by antibiotics, especially streptomycin. Rodent control has advanced enormously, chronic rodenticides like warfarin largely replacing acute poisons. Modern synthetic insecticides, especially DDT, remain an important weapon for dealing with plague epidemics. Perhaps I should conclude with a cautionary note, by mentioning that warfarin-resistant strains of rats and DDT-resistant strains of fleas have appeared, so that we may not be able to rely on these weapons indefinitely.

The Elucidation of Typhus

In contrast to the voluminous early works on plague, comparatively few historical accounts of typhus can be recognized with certainty. It did not stand out so starkly, because it was prevalent under conditions conducive to various other diseases caused by bad sanitation and insufficient food. It was a disease associated with besieging armies, beleaguered cities and prisons. Under these circumstances, specific diseases were not always distinguishable with clarity, nor were there often intelligent and dispassionate physicians available to record signs and symptoms. By the sixteenth century, however, typhus epidemics were definitely recognizable in military campaigns. But it was not exclusively confined to warfare and it may have been more familiar to the less adventurous physicians as jail fever. Not unnaturally, they concluded that it was caused by the stench of prisons and prisoners. Francis Bacon[16] (1558–1601) gave some opinion of this kind, noting, however, that: 'not those stinks which the nostrils abhor . . . that are the most pernicious, but such airs as have similitude with a man's body, and so insinuate themselves and betray the Spirits.' George Wither[323] (1588–1667) the poet and pamphleteer, likewise absolved foul stinks.

> It was no noysome *Ayre*, no *Sewre*, or *Stinke*
> Which brought this *Death*, as most among us thinke
> For then these places where ill smells abound
> Had more infection at that time been found

> Than we perceive there were: yea this *Disease*
> Upon each person delicate would sieze
> Without exception.

Likewise, he dismissed overcrowding as a cause of plague and affirms a belief in its infectiousness.

> ... this *Plague* Ev'n naturally affects
> a space of aire about it, and infects
> (At such and such a distance) ev'ryone
> As he hath weakness, to worke upon.

But, he concludes, not everybody is equally susceptible.

> Witness the survival of the Clarkes
> The *Sextons, Searchers,* Keepers and those sharks
> The shameless *Bearers* ...
> How scap't the *Surgeon,* That oft put his head
> Within the steame of an Infectious bed.

John Gadbury,[98] the astrologer, was so impressed by the apparent immunity of some individuals that he was sceptical of the possibility of infection. In any case, he points out, the very *first* case of an epidemic cannot result from infection; it must be '*ex Astris*'; and if the first case, why not the second, third, thousandth or millionth? The low standards of hygiene in the seventeenth century must, indeed, have cast some doubts on the power of stinks, which were so general. Gadbury ridicules those who held their noses when passing a plague house yet would confer closely with unsavoury friends 'Nor *think evilly* of *lying* with a *Husband* or *Wife,* whose *breaths* or *issues* (for wholesomness) are many degrees *below Carion,* a *Jakes* or a *Charnel-house.*'

While the wretched prisoners were the only victims of typhus, it did not command much attention; but occasionally it managed to spread and involve other members of criminal courts. MacArthur[190] described a sensational outbreak following an assize at Oxford in 1577, when two judges, a sheriff and under-sheriff, six J.P.s and most of the jury died. Within twelve days, a hundred members of the University had followed them and all in all, the death roll was 510. The Black Assize at Oxford was by no means the only case of its kind; various others could be cited, notably one at the Old Bailey in 1750. On this occasion, the Lord Mayor, and other notables on the bench lost their lives, as well as 40 other members of the court. This caused considerable concern, and the Aldermen and Corporation of the City appointed a Board of Enquiry to consider the cause of the trouble. This committee reached the unoriginal conclusion that the infection was due to the stink of the prisoners and that improved ventilation was desirable. A remarkable device for extracting foul air from Newgate prison, operated by a kind of windmill, was erected some time later. In the course of its construction, seven of the eleven workmen employed contracted jail fever! As a practical measure, it could be put on a par with a nosegay which used to be carried by judges to guard them from the effluvia of the court.

It was about this time that Dr Johnson observed that 'Being in a ship is being in jail, with the chance of being drowned . . . a man in jail has more room, better food and commonly better company'. It is therefore no surprise to find jail fever rife in the eighteenth century navy. James Lind (1716–94) describes many such cases in his *Two Papers on Fevers and Infection*[186] (1763). At one point, he remarks:

. . . what was very shocking, a great many of the patients from the *Portmahon* whom I visited, had never changed their clothes, from the time they were pressed in *June* to the 22nd of *October*, when they came to the hospital. The unclean linen and rags which they had lain in for about four months were sufficient to have bred infection.

Lind had other sound observations on the subject of patients' clothing, noting for example that many nurses 'who were tainted, became so by keeping the dirty linen after it had been taken from the sick, for some days in the rooms where they slept, contrary to the rules of the house'. In general, however, he still imagined the contagion to be a kind of noxious effluvium and describes various methods of fumigating ships by burning charcoal or brimstone. In this respect, Lind was in agreement with his eminent contemporary, Sir John Pringle (1707–82) who was concerned with *Diseases in the Army in Camp and Barracks*[252] (1753). Pringle recognized that 'the symptoms of the jail fever were in every point so like those of hospital fever that, as they were conjectured to be the same before, they were now proved to be really such.' He puts the cause down to '. . . filth and impurity and poisonous *effluvia* or sores, mortification, dysenterie and other septic excrements'.

Reading the signs and symptoms of jail or hospital fever set out by Pringle, one may suspect that only some of the patients were actually suffering from typhus. It was not until the nineteenth century that typhus was recognized as a distinct entity. The name is derived from the Greek '*typhos*' meaning hazy and was used by Hippocrates to convey a confused state of mind with a tendency to stupour. The word was first applied to this disease by the French doctor and philosopher Sauvages de la Croix[276] (1706–67) who attempted to classify diseases 'in the way botanists classified plants'. It is indeed very apt, since typhus causes a peculiar drunken stuperose look, not seen in any other disease, except perhaps plague. Nevertheless, various other diseases cause a degree of stupor and Sauvages included about a dozen under the generic heading TYPHUS, though, indeed, he gives true typhus (*T. septica* or jail fever) as the first type. A common confusion was with the enteric disease typhoid, as the similar name implies. A categorical distinction was claimed by the American doctor, W. W. Gerhard,[108] who had seen both forms in Paris; the enteric infection among the inhabitants and the tickettsial disease in soldiers returning from the wars. In later observation of typhus in Philadelphia (1836) he speculated on the mode of transmission, which he concluded was by contagion or infection (e.g. exhaled breath).

Relapsing fever was also liable to be considered as a variant of typhus, especially as it was liable to occur under the same conditions, and an early German name for it was *Rückfalltyphus*. This error began to be realized in the first half of the nineteenth century, largely by the observations of Professor W. Henderson in Edinburgh, so that by 1862, Charles Murchison[220] (1830–79) was able to write (in his Treatise on the Continued Fevers of Great Britain):

The investigations of Henderson and other writers on the epidemic of 1843, established the specific distinctness of relapsing fever from typhus, while those of Gerhard, Stewart, Jenner and others have proved the non-identity of true typhus and the "typhoid fever", so ably described by Louis.'

Nevertheless, there was a long way to go before this outlook became general, especially as many physicians believed that diseases could transform themselves, one into another, as a result of changed conditions. For example, as late as 1890, a noted authority, Charles Creighton, in his *History of Epidemics in Great Britain*[71] could write:

Yellow fever is a typhus of the soil, whereas the common and much less fatal typhus of ordinary domestic life in colder lattitudes is an infection above ground—of the air, water, floors and furnishings of rooms. There is the same relation between yellow fever and ordinary typhus in that respect, as between plague and ordinary typhus. When ordinary typhus as passed into a soil poison by aggravation of conditions, it becomes at the same time, bubonic fever or, approximately, plague proper.

Despite such relapses, the modern idea of typhus gradually prevailed;

and in the latter part of the century, the suspicion grew that typhus might be transmitted by insect parasites. According to Zdrodovskii and Golinevich,[329] it was a Russian scientist, Gregorey Nikolayevich Minkh, who in 1878 suggested that it might be spread from man to man 'by means of blood-sucking insects.' Nicolle, however, noting that the association of typhus with poverty and squalor was well known, wrote (in 1910), 'Without affirming that the parasites of body or clothes are the agents of transmission, several authors have put forward suggestive opinions.'[221] Especially, he noted, MM. Netter and Thoinot in their *Rapport General sur le typhus en France de 1892 à 1893*. These suggestions accorded with Nicolle's own observations in North Africa, where he observed epidemics of typhus which mainly affected the natives who were often lousy but not Europeans who were free of lice, though often bitten by fleas, etc. Charles Nicolle (1866–1936) was a leading member of a brilliant team at the Institut Pasteur of Tunis. As intern at a Paris hospital, he was afflicted with deafness and this turned his career from clinical practice to research. Although he worked on other tropical diseases with success, it was his work on typhus for which he was best known and which resulted in his Nobel Prize award in 1928. The first step was to induce infections in laboratory animals; first monkeys and later other animals. By 1910, he and his collaborators had succeeded in transmitting typhus from infected macaque monkeys to healthy ones by transferring to them human lice which had been fed on the diseased animals.

The rickettsial organisms responsible for typhus had not yet been specifically identified; but the young American, Howard Taylor Ricketts (1894–1910) had found related forms in patients with Rocky Mountain spotted fever and suspected that similar organisms were responsible for typhus, when he contracted the disease and died in Mexico.

The problem was then pursued by both Henriques Da Rocha Lima (1880–?), a Brazilian doctor, and Stanislaus von Prowazek (1875–1915) an Austrian biologist. Von Prowazek had made extensive researches on micro-organisms, especially protozoa and perhaps by association with Ehrlich at Frankfurt and later with Fritz Schaudinn, he latterly turned his attention to medical parasites. In 1914, Rocha Lima came to Europe and was sent in the company of von Prowazek to Constantinople to study typhus in Turkish prison camps connected with the Balkan Wars. After the outbreak of the World War in 1914, the two returned to the *Institut für Schiffs- und Tropenkrankheit* at Hamburg and thence to further studies of typhus among Russian prisoners. As a result, von Prowazek contracted the disease and died early in 1915. Rocha Lima continued his studies and in 1916 named the pathogen *R. prowazeki* in honour of his collaborator. At the same time, the German authorities became aware of the importance of lice and studies of their biology and control were undertaken, notably by Albrecht Hase.

On the Western Front, the Allies were fortunately spared the presence of typhus, but were hampered by a less malignant louse-borne rickettsial disease, Trench fever. Accordingly, studies of louse biology and control were energetically done in England, largely by G. H. F. Nuttall (1862–1937) at Cambridge and Arthur W. Bacot at the Lister Institute in London. The two men presented a great contrast. Nuttall, of American birth, had travelled widely and studied at Göttingen as well as in Baltimore. Medically and biologically trained, he spoke several languages and ended as a Cambridge Professor. Bacot was a simpler, less cultured man, who worked in a business firm in the City from the age of 16 to 44, during which time, however, he was an energetic amateur entomologist. Between 1893 and 1909, he published more than fifty papers, mainly on the lepidoptera. In 1909, he was invited by his friend Major Greenwood (1880–1949) an eminent medical epidemiologist, to assist in studies of fleas in relation to plague, at the Lister Institute. I have already mentioned the successful outcome of some of this work, due to his untiring zeal and skill. From fleas, he turned to lice and to work on trench fever, in collaboration with microbiologists at the Lister. After the war, Bacot continued to work on lice and louse-borne diseases. In 1920, he was invited to join the U.S. Red Cross expedition to Poland to investigate typhus. There he met the ingenious Rudolf Weigl (1883–1957) who was rearing quantities of lice and infecting them with rickettsiae through the anus with a very fine pipette. Bacot acquired this technique and practised it on his return to London, where it was especially valuable since immune typhus convalescents were not available. Further researches with E. E. Atkin[4] established further important details of typhus transmission; notably that it was through the faeces of the louse (scratched into small abrasions or inhaled) rather than its bite which caused infection. In November, 1921, Bacot was invited to study typhus in Cairo by the Egyptian Government and he went early in 1922. Soon after arrival, however, he contracted the disease and he died on 12 April.

The volume of research on lice and louse-borne diseases subsided to some extent in between the two World Wars; but a spate of papers appeared on various aspects during the 1940s. The complexities of these modern researches are too extensive and specialized for this book. Most of the wartime work on lice is summarized in the 1945 edition of Buxton's book *The Louse*.[54] Concurrent studies of the pathogen can be found in textbooks of rickettsial diseases. In brief, the advent of DDT enormously simplified the suppression of typhus epidemics. New methods of culturing rickettsiae on a large scale were developed to provide anti-typhus vaccine. The laborious method of Weigl was replaced by accumulation in the lungs of infected rodents, in tissue culture or in egg yolk.

Advances since the war can be found in the PAHO/WHO symposium on *The Control of Lice and Louse-borne Diseases*,[230] published in 1973. Vaccines remain a useful prophylactic for sanitary personnel who have to cope with epidemics and antibiotics have been introduced for curing patients. Insecticide powders are still the main weapon to quell epidemics by general de-lousing. Unfortunately, in many parts of the world, lice show signs of resistance to DDT and other chlorinated insecticides; and in an epidemic in Central Africa in 1972, the lice were found to be resistant to malathion, which is generally regarded as an effective alternative to DDT.

The Elucidation of Louse-borne Relapsing Fever

It is difficult to trace with any certainty very ancient records of relapsing fever. An epidemic in the island Thassus, off Thrace, which was described by Hippocrates, may have been this disease. It seems, however, that the earliest reliable records date from the eighteenth century as noted by Murchison[220] in 1862; perhaps the first was an epidemic in Dublin in 1739, recounted in *Chronological History of the Diseases of Dublin* by a Dr Rutty. From the fact that both diseases were liable to occur among louse-infected people under conditions of poverty and deprivation, it was generally thought that both were variants of the same disease. Doubts began to arise in the earlier decades of the nineteenth century. It was noticed that the relapsing type of fever was more prevalent earlier in epidemics; and the non-relapsing, but more often fatal, true typhus became dominant later. A careful evaluation of the differences was made by the Scottish doctor, William Henderson[125] (1810–72) after observing an epidemic in Edinburgh in 1843, which was almost exclusively relapsing fever. He noted the common belief in the mutability of diseases held by many medical men at the time, quoting a Dr March of Dublin who 'has known typhus produced by the contagion of smallpox and intermittent fever by the contagion of typhus'. Likewise, a Dr Southwood Smith 'virtually affirming that intermittent, remittent, typhus fever and plague all originate from different intensities of the same poison.' Henderson himself, however, claimed to have extensive experience of typhus, amounting to some 1600 to 2000 cases. After considering the characteristic signs and symptoms of the recent epidemic, he decides that it is of a distinctly different nature from typhus; nor did he believe that an infection of the one could arise from contagion with the other. Henderson later became converted to homeopathy which seems a sad lapse in such a careful observer. The categorical distinctness of relapsing fever was upheld by Murchison,[220] who gave a number of sound reasons why he believed that both diseases were normally spread by contagion. On the other hand, he was inclined to suspect that both could occasionally be generated spontaneously under conditions conducive to the accumulation of putrid poisons.

The spirochaetes responsible for louse-borne (or 'European') relapsing fever were discovered in the blood of patients by Otto Obermeier[227] (1843–73) and demonstrated to the Berlin Medical Society early in 1873. He was also interested in cholera, and, later that year, he contracted that disease and died. Obermeier had observed that the 'spirillae' were plentiful in the blood during the attacks of fever and disappeared in between the relapses; but further work was hampered by various difficulties. The spirochaetes could not be cultured *in vitro* (this was not achieved until 1912, by Hideyo Noguchi,[223] working in New York; and then not in permanent culture). A step forward was made when Koch and H. V. Carter independently showed that monkeys could be infected.

There remained the problem of transmission, for which clues were provided by discoveries of other insect-borne diseases at the turn of the century. According to Nuttall,[225] the first to suggest transmission of relapsing fever was Flügge, in 1891. Presumably this was Carl G. F. W. Flügge; but I have not been able to trace the reference. In 1897 the Russian J. Tictin[299] published observations of a relapsing fever epidemic in Odessa. He noticed that the disease flourished among sailors inhabiting rough lodging houses near the docks, where they slept in rags beset with bugs, lice and fleas. Suspecting these parasites, he examined their stomach contents and found spirochaetes in the bed bugs. By collecting blood from bugs fed on fever patients (up to 48 hours previously) he was able to induce relapsing fever in monkeys injected with this blood. He concluded that bugs were the vectors, either by their contaminated mouthparts or by their infected faeces, scratched into the skin. In 1902, Justin Karlinski[164] came to similar conclusions in observing epidemics in the Balkans. According to him, the spirochaetes remained viable for a month in the bed bug gut. Shortly after this, C. Christie[61] in India attempted to transmit the disease to himself by bites of bugs from infested houses with fever patients; and in Liverpool, Breinl *et al.*[35] attempted similar transmission tests with monkeys. Both trials were negative and later work has excluded bed bugs from any important contribution to relapsing fever epidemics.

Suspicion gradually turned to lice. Captain Percival Mackie,[193] of the Indian Medical Service, working in Bombay made some suggestive observations at an epidemic at a children's home in 1909. The disease was much more prevalent among the boys, among whom body lice were common, than among the sparsely infested girls. The sleeping quarters of both sexes were, however, equally infested with bed bugs. He concluded that lice were the vectors and demonstrated spirochaetes in the stomach of recently fed specimens.

The time was ripe for the final attack on the problem, which was made by French workers at Pasteur Institutes in N. Africa. Sergent and Foley,[282] working in Algeria, considered the possible role of ticks, mosquitoes, bugs, lice, fleas and flies as vectors, and on various grounds, decided that lice were most probable, They describe some transmission experiments using lice in attempts to transmit the disease from man to monkey or man to man by bites or by wearing infested clothing. The results generally supported the hypothesis, though transmission was capricious and never by simple louse bites. The final discoveries were published in 1913 by C. Nicolle and his colleagues in Tunis.[222] They showed that the spirochaetes disappeared from the louse gut to reappear in the body cavity, but never in the mouthparts. The conclusion was the infection would only result on bursting of infected lice near an abrasion or the conjunctiva of the eye.

Subsequent researches in the twentieth century have investigated the reason for relapses, which seem to be due to waves of antibody produced by the infected animal or man. Between relapses, the spirochaetes persist in small numbers in the brain and other organs. Some improvements have been made in culture *in vitro* and the organisms can be readily grown in chick embryos. For treatment, antibiotics, especially chloramphenicol are prescribed; but mass delousing is the most important means of combating epidemics, as in the case of typhus. Further work on lice up to 1945 was summarized by Buxton[54] and brought up to date by a 1972 PAHO/WHO Seminar.[230]

References

1 'A. B.', (A letter to Mr Urban) *Gentleman's Magazine*, (p. 534, October 1746).
2 Adams, Joseph (1756–1818), *Observations on Morbid Poisons, Chronic and Acute*, London (1807).
3 Aesop (619–564 B.C.), *The Fables . . . as first printed by Caxton in 1484*, (ed. Joseph Jacobs) London (1889).
4 Aitkin, E. E. and Bacot, A. W., (Transmission of typhus by lice) *Brit, J. exp. Path.*, (1922) **3**, 196.
5 Albertus Magnus (1193–1280), *De Animalibus*, (p. 252 verso) Venice (1495).
6 Aldrovandi, Ulysses (1522–1605), *De Animalibus Insectis*, Bologna (1602).
7 —, *Monstrorum Historia* (etc.), Bologna (1642–8).
8 Alt, H. Ch., *Dissertation de Phthiriasi*, Bonn (1824).
9 Altman, P. L. and Ditmer, D. S. (ed.), *Biology Data Dook*, Fed. Amer. Soc. Exp. Biol., Bethesda, Md. (1972).
10 Amatus Lucitanus (Joao Roderiguez, 1511–68), *Curationum medininatum centuria quatuor*, (p. 274) Basle (1556).
11 Andersen, Hans C. (1805–75), *Fairy Tales*, Odense and London (1960).
12 Anon. (1951), 'Londoner's Diary', *Evening Standard* (5 March 1951).
13 Appian of Alexandra (Second century A.D.), *The Civil Wars*, (trans. H. White) Oxford (1913).
14 Aristophanes (448–? B.C.), *Clouds*, (trans. Hickie) London (1853); *Frogs*, (trans. Rogers) London (1924).
15 Aristotle (384–322 B.C.), *Historium Animalium*, (trans. J. Smith and W. Ross) Oxford (1910); *De Generatione Animalium*, (trans. Smith and Ross) Oxford (1912); *Problemata*, (trans. E. S. Foster) Oxford (1927).
16 Bacon, Francis, (1561–1626), *Sylva Sylvarum*, London (1627).
17 Bacot, A. (Bionomics of Fleas) *J. Hygiene*, (Plague Supplement iii) Cambridge (1914).
18 —, (Biology of Lice) *Parasitol.*, (1917). **9**, 228.
19 Bartholomeus Anglicus (fl. 1230–50), *De Proprietibus Rerum*, (English edition) London (1535).
20 Bateman, Thomas (1778–1821), *Practical Synopsis of Cutaneous Diseases According to Dr Willan*, London (1813).
21 Bayle, Pierre (1647–1706), *Dictionaire Historique et Critique*, Paris (1695–7).
22 Beaumont, F. (1584–1616) and Fletcher, J. (1579–1625), *Thierry and Theodora*, (Beaumont and Fletcher, vol. X) Cambridge (1912).

23 Benjamini, E. *et al.*, (Flea Bite Reactions) *Expt Parasitol.*, (1963) **13**, 143.
24 Bennet-Clark, H. C. and Lucey, E. C. A., (Energetics of Jumping Fleas) *J. exp. Biol.*, (1967) **47**, 59.
25 Berthold, A. A. (1803–61), *Lehrbuch der Zoologie*, (Phthiriasis; pp. 349 and 353) Göttingen (1845).
26 Blacklock, B., (Tests of poisons against bed bugs) *Ann. trop. Med, Parasit.*, (1912) **6**, 415.
27 Bodenheimer, F. S., *Materialen zur Geschichte der Entomologie bis Linne*, Berlin (1928–9).
28 Boehn, M. von, *Modes and Manners*, (trans. Joshua) London (1932).
29 Bogdandy, S., (Trapping bed bugs in Balkans) *Naturwiss.*, (1927) **15**, 474.
30 Boghurst, William (1631–85), 'Loimographia, or an HistoricalAccountof the Last Plague in London (1666)', (ed. J. F. Payne) *Supplement trans. Epidem. Soc, London*, (1894) **13**, 1.
31 Bolam, R. M. and Burtt, E. T., 'Flea infestation as a cause of papular urticaria,' *Brit. med. J.*, (1956) **i**, 1130.
32 Borel, Pierre (Petrus Borellus, *c.* 1620–87), *De vero telescopii inventore ... Accessit etiam centuria observationem microcospicarum*, Hague (1655).
33 Bossche, G. van Den (fl. 1638), *Historia Medica*, Brussels (1639).
34 Brehm, A. E. (1829–84), *Tierleben*, (Bd. II) Leipzig and Vienna (1892).
35 Breinl, A., Kinghorn, A. and Todd, S. L., (Attempts to transmit spirochaetes by bed bugs) *Mem. Liverpool Sch. trop. Med.*, (1906) (21) 113.
36 Bremser, J. G., (1767–1827), *Über lebende Würmer im lebenden Menschen*, (lousy disease, pp. 54–5) Vienna (1819).
37 Bruckmann, F. E. (1697–1753), *Die neuerfundene curiose Flöh-Falle*, Wolfenbüttel (1729).
38 Brues, C. T., 'Animal life in hot springs', *Quar. Rev' Biol.*, (1927) **2**, 81.
39 —, *Insect Dietary*, Harvard (1946).
40 Buchan, William (1729–1805), *Domestic Medicine; or the Family Physician*, Edinburgh (1769).
41 Burckhardt, J. (1818–97), *Civilization of the Renaissance in Italy*, (1860; trans. Middlemore, 1878) London (1944).
42 Burmeister, H. (1807–?), *Handbuch der Entomologie*, (Phthiriasis: Vol. 1 pp. 331–94) (1832).
43 Burns, Robert (1759–96), *Poetical Works*, (ed. J. L. Robertson) London (1896).
44 Busch, G., (Entomophobia) *Angew. Parasitol.*, (1960) **1**, 65.
45 Busch, Wilhelm (1832–1908), *Julchen*, Munich (1883).
46 Busvine, J. R., 'Relative Toxicity of Insecticides', *Nature*, (1942) **150**, 208.
47 —, 'Simple Experiments on the behaviour of Body Lice', *Proc. R. ent. Soc. Lond. (A)*, (1944) **19**, 22.
48 —, (Control of lice by insecticides prior to DDT) *Bull. ent. Res.*, (1945) **36**, 23–32.

49 —, 'The Head and Body races of Lice', *Parasitol.*, (1948) **39**, 1.
50 —, 'Analysis of 685 enquiries regarding insect pests of hygienic importance', *Mon. Bull. Min. Hlth and Pub. Hlth Lab. Service*, (1955) **14**, 178.
51 —, 'Progress in the eradication of bed bugs', *Sanitarian*, (1957) **65**, 365.
52 —, *Insects and Hygiene*, (pp. 59 and 60) (2nd edn.) London (1966).
53 Buxton, P. A., *Animal Life in Deserts*, London (1923).
54 —, *The Louse*, London (1939).
55 Cardano, Girolamo (1501–76), *De rerum varietate et de subtilitate rerum*, (in *Opera Omnia*) Leyden (1663).
56 Catullus, Gius Valerius (?84–54 B.C.), *Poems*, (ed. Merrill) Lipsiae (1923).
57 Celsus (fl. 30 A.D.), *De Medicina*, (trans. Spencer) London and Harvard (1935–8).
58 Cestoni, D. J. (1637–1718), 'Letter concerning Redi's mss and life cycle of the flea'. *Phil. Trans. Roy. Soc.*, (1699) **21**, 42.
59 Chapman, T., 'The Future of Pest Control', *S.P.A.N.*, (1973) **16**, 51.
60 Chaucer, Geoffrey (1340–1400), *Works*, (ed. W. W. Skeat) Oxford (1903).
61 Christie, C., (Transmission of relapsing fever: bugs suspected) *J. trop. Med.*, (1902) **5**, 39.
62 Christophers, R., 'A zoological view of malaria', *Proc. R. Soc. Med.*, (1934) **27**, 991.
63 Churchill, A. and Churchill, J., *Collection of Voyages and Travels*, London (1732).
64 Clay, Theresa, 'Biting Lice of Birds', Symposium, *Host Specificity among Parasites of Vertebrates*, Univ. Neuchatel (1957).
65 Cockburn, A., *The Evolution and Eradication of Infectious Disease*, Baltimore (1963).
66 Conway, M. D., *George Washington's Rules of Civility*, London (1890).
67 Corn, M., (Adhesion of Particles) in *Aerosol Science* (ed. Davies) London (1966).
68 Cowan, F., *Curious Facts in the History of Insects*, Philadelphia (1865).
69 Cowper, W. (1731–1800), *Poetical Works*, (ed. S. Milford) Oxford (1905).
70 Cragg, F. W., 'Fertilization in the bed bug', *Indian J. med. Res.*, (1915) **2**, 698; ibid., (1920) **8**, 32.
71 Creighton, C., *A History of Epidemics in Britain*, Cambridge (1894).
72 Culpepper, G. H., (Rearing human lice on rabbits), *Am. J. trop. Med.*, (1948) **28**, 499.
73 Darwin, Charles, *The Voyage of the 'Beagle'*, London (1905).
74 Descartes, Rene (1596–1650), *Prima cogitations circa Generationem animalium*, (in *Oeuvres*, ed. C. A. and P. Tannery) Paris (1909).
75 D'Israeli, I. (1834) *Curiosities of Literature*, (I. 132) London (1834).
76 Diodorus of Sicily (First Century B.C.), *Works*, (trans. Oldfather) London (1935).
77 Dioscorides, Pedanius (fl. 54–68), *Materia Medica*, London 1655, (trans. J. Goodyear) Oxford (1933).

78 Dobell, C., *Antony van Leeuwenhoek and His 'little Animals'*, London (1932).

79 Donne, John (1573–1631), *Poems*, (ed. Grierson) Oxford (1912).

80 Durr, G. E. F., (Origin of lice in phthiriasis) *J. pract. Heilk.* (1840) **90,** 89.

81 Elloposcleros (Pseud. Johann Fischart) (1545–1614), *Flöh Hatz, Weibe Kratz, etc.*, Strassburg (1601).

82 Evans, G. W., Sheals, J. G. and McFarlane, D., *Mites of the British Isles,* British Mus. (Nat. Hist.), London (1961).

83 Eysell, Adolph, (Chinese flea trap) *Handbuch der Tropenkrankheiten,* (1913) **1,** 87.

84 Fain, A., 'Variability and affinities of *Sarcoptes scabiei'*, *Acta Zool. Path. Antwerp.,* (1968). No. 47.

85 —, Adaptation to parasitism in mites', *Acarologia*, (1969) **11,** 429.

86 Fallen, C. F. (fl. 1800–30), *Monograph Cimicum Sueciae*, Hafniae, (1807).

87 Farrant, J. and Smith, A., 'Low temperature biology', *Penguin Science Survey B*, p. 63 Harmondsworth (1966).

88 Ferris, G. F., 'The Sucking Lice', *Memoires of Pacific Coast Ent. Soc.,* San Francisco (1951).

89 Feytaud, J., 'Les Puces dans le Folklore et la litterature', *Actes Acad. Sci. Belles-lettres et Arts de Bordeaux.,* (1970) 4e ser'. 24.

90 Finch, A. T., *Ancient Church at Clere now called Kingsclere*, Winchester (1905).

91 Folsom, J. W. and Wardle, R. A., *Entomology and Ecology*, (4th edn.) London (1934).

92 La Fontayne, (1621–95), *Oeuvres Completes*, Paris (1965).

93 Forbes, J., *Oriental Memoires*, London (1813).

94 Frascator, Jerome (1478–1553), *De Contageone*, Venice (1546).

95 Friedman, R., 'The Story of Scabies', *Medical Life*, (1934) **41,** 381, 426, 620; (1935) **42,** 218; (1936) **43,** 167.

96 Fritsch, W., 'The mite genus *Harpyrhynchus'*, *Zool. Anz.,* (1954) **152,** 177.

97 Fuchs, C. H. (1803–55), *Die Krankhaften Veränderungen der Haut,* Göttingen (1840).

98 Gadbury, John (1627–1704), *London's Deliverance Predicted*, London (1665).

99 Galen (121–201), *Opera Galeni,* (ed. D. C. G. Kühn) Leipzig (1826).

100 Garnham, P. C. C., *Malaria Parasites*, Oxford (1966).

101 Gascoigne, R. (1537–77), *Works*, (ed. Cunliffe) (1907–10).

102 Gaulke, (Cases of lousy disease) *Vjschr. gerichtl. Med.,* (1863) **23,** 315.

103 —, 'Uber die Lausesucht (Phthiriasis)', *Wien, med. Woschr.,* (1866) **16,** 380 and 398.

104 Gay, John (1685–1732), *Works*, (Chap. 4) London (1775–7).

105 Geer, C. de (1720–78), *Memoires pour Servir a L'histoire des Insectes,* (8 vol.) Stockholm (1752–8).

106 Geoffroy, E. L. (1725–1810), *Histoire Abregèe des Insectes qui se Trouvent aux Environs de Paris*, Paris (1762).

107 Geoponika, (Agricultural and Domestic Pursuits) (trans. from Greek, T. Owen) London (1806).

108 Gerhard, W. E. (1809–72), 'On the typhus fever', *Am. J. med. Sci.*, (1837) **19**, 289.

109 Godfrey, Boyle (d. 1756?), *Miscellanea vere utilia*, London (1737).

110 Goethe (1749–1822), *Faust*, (trans, T. Martin) Edinburgh and London (1872).

111 Grego, Joseph, *Rowlandson the Caricaturist*, London (1880).

112 Grey, Elizabeth (Countess of Kent) (1581–1651), *Rare Secrets of Physicke*, London (1653).

113 Grimmelshausen, H. I. (1625–76), *Simplicissimus*, 1669 (English translation), London (1912).

114 Guy de Chauliac (1300–68), *Chirugia magna*, (Anon. English translation, sixteenth century) (ed. M. Ogden) London (1971).

115 Hardy, D. E., Introduction and background information, *Genetic Mechanisms of Speciation in Insects*, (ed. M. J. D. White) Sydney (1974).

116 Harvey, William (1578–1657), *Anatomical Exercitations concerning the Generation of Animals*, London (1653).

117 Hase, A., (Biology of lice) *Zeit. angew. Entom.*, (1915) **2**, 265; *Naturwiss. Woch. schr. Folge*, (1916) **15**, 1; *Zbl. Bakt. 1 Abt. Orig.*, (1919) **82**, 461.

118 ——, (Biology of the bed bug) *Z. angew. Ent.*, (1917) **4** (Beiheft l); *Z. Parasitenk.*, (1930) **2**, 368 ibid., (1931) **3**, 837.

119 Hawkins, Francis, *Youths Behaviour, or Decency in Conversation among Men*, London (1646).

120 Herberden, W. (1710–1801), *Commentary on the History and Cure of Diseases*, London (1802).

121 Hebra, H. (1816–80), *Diseases of the Skin*, (trans. C. H. Fagge and P. H. Pye-Smith) London (1868).

122 ——, (Phthiriasis controversy) *Wien. med. Wschr.*, (1866) **16**, 425, 441 and 457.

123 Heilesen, B., *Studies on Acarus scabiei and Scabies*, Copenhagen (1946).

124 Helmont, J. B. van (1577–1644), *Oriatrike, or Physick Refined*, (trans. J. Chandler) London (1662).

125 Henderson, W. (1810–72), (Differentiation of relapsing fever from typhus) *Edinb. Med. & Surg. J.*, (1844) **61**.

126 Heracleitus (Sixth Century B.C.), *Fragment 56 (Ancilla to the pre-Socratic Philosophers)*, (ed. K. Freeman) Oxford (1948).

127 Herbert, Anne (1590–1676), *Diary of the Lady Anne Clifford*, (ed. Sackville-West) London (1923).

128 Herodotus (480–408 B.C.), *History*, (trans. G. Rawlinson, ed. Blakeney) London (1912).

129 Herrlinger, G., *Totenklage um Tiere in der Antiken Dichtung*, Stuttgart (1930).

130 Heusinger, K. F. (1792–1883), *Recherches de Pathologie comparée*, (Phthiriasis, Vol. 2 (3) 594) Cassel (1847).

131 Hildegarde of Bingham (1099–1179), (Animal book) (trans. to German by Huber) Vienna (c. 1923).

248   *References*

132 Hill, Thomas, *Natural and Artificial Conclusions*, (2nd edn.) London (1650).
133 Hinton, H. E., 'The panorpoid Complex', *Ann. Rev. Entom.*, (1958) **3**, 181.
134 —, (Thermal tolerance of chironomid larvae) *Nature*, (1960) **188**, 336.
135 Hippocrates (460–357 B.C.), *Oeuvres*, (trans. to French, E. Littre) Paris (1839–61).
136 Hirst, L. F., *The Conquest of Plague*, Oxford (1953).
137 Hoare, C. A., 'Evolution of mammalian trypanosomes', *Advances Parasitology*, (1967) **5**, 47.
138 Hocking, K., 'Louse control through textile fibre size', *Bull entom. Res.*, (1957) **48**, 507.
139 Hoeppli, R. and Ch'iang, I-H., 'Insect Vermin in Ancient Chinese literature', *Chin. med. J.*, (1940) **58**, 338.
140 Hoffman, E. T. W. (1776–1822), *Meister Floh*, (in *Späte Aerke)* Munich (1965).
141 Holland, G. P., 'Evolution and Classification of Fleas', *Ann. Rev. Entom.*, (1964) **9**, 123.
142 —, 'Fleas of New Guinea', *Mem. entom. Soc. Canad.*, (1969) No. 61.
143 Hooke, Robert (1635–1783), *Micrographia*, London (1665).
144 —, *Lectures and Collections*, London (1678).
145 Hopkins, G. H. E., 'Host relations of lice of mammals', *Proc. Zool. Soc. London*, (1949) **119**, 387.
146 —, (Host associations of fleas (pp. 64–87) and lice (pp. 88–119)) Symposium, *Host Specificity among Parasites of Vertebrates*, Univ. Neuchatel (1957).
147 Hopkins, G. H. E. and Rothschild, Miriam, *Catalogue of Fleas in the British Museum*, Vol. I. B. M. (Nat. Hist.), London (1953).
148 Horace (Quintus Horatius Flaccus) (65–8 B.C.), *Satires*, (trans. Fairclough) London (1926).
149 Hortus Sanitatis, (Facsimile of Translation by Laurence Andrewe) (fl. 1510–37) (ed. Noel Hudson) London (1954).
150 Hughes, T. E., *Mites or the Acari*, London (1959).
151 Humphries, D. A., (Combs of fleas; water drinking; movements in cocoon) *Entomologists Mon. Mag.*, (1966) **102**, 232 and 260; *Proc. R. ent. Soc. Lond.*, (1967) **42**, 62.
152 —, (Mating in the hen flea) *Proc. R. ent. Soc. Lond. A.*, (1967) **42**, 101; *Animal Behav.*, (1967) **15**, 82.
153 Hunter, John (1728–93), *Observations on the Diseases of the Army in Jamaica*, London (1788).
154 Huxley, Julian, 'Sizes of living things', *Uniqueness of Man*, London (1941).
155 Janisch, E., (Biology of the bed bug) *Z. Parasitenk.*, (1933) **5**, 460; ibid., (1935) **7**, 408.
156 Jeitteles, A. L. J. (1799–1878), (Phthiriasis) *Oesterr. med. Wschr. Fasc.*, (1841) **2**, 632.
157 Jenkins, J. S., *Voyage of a U.S. Exploring Squadron 1838–42*, Auburn,

U.S.A. (1852).
158 Joerdens, J. H. (1764–1813), *Entomologie und Helminthologie des menschlichen Körpers.*, Hof, G. A. Grau (1801).
159 John XXI (Pope) (d. 1276), *The Treasury of Health,* etc. (trans. Humphry Llwyd, 1527–68) London (*c.* 1550).
160 Johnson, C. G., Ecology of the bed bug, *J. Hyg.,* (1941) **41,** 345.
161 Jonstone, John (1603–75), *Thaumetographia naturalis,* Amsterdam (1632).
162 Julian, Emperor (The Apostate) (331–63), *Misopogon,* (trans. Wright) London (1913).
163 Kalm, Peter (1715–79), *Travels into North America,* (trans. J. R. Foster) Warrington (1770).
164 Karlinski, J., (Transmission of relapsing fever: bugs suspected) *Zentbl. Bakt.,* (1902) **31,** 566.
165 Kelin, D. and Nuttall, G. H. F., (Relations of head- and body-lice of man) *Parasitol.,* (1919) **11,** 279.
166 Kemper, H., (Biology of the bed bug) *Zeit. Morph. Oekol. Tiere,* (1930) **19,** 160 ibid., (1932) **24,** 491.
167 —, Die Bettwanze, *Klientiere and Pelztiere,* (1936) **12.**
168 Kiel, H., The Louse in Ancient Greece, *Bull. Hist. Med.,* (1951) **25,** 305.
169 Kirby, W. and Spence, W., *Introduction to Entomology,* London (1815).
170 Kircher, Athanasius (1601–80), *Scrutinium Physico-Medicorum Pestis,* Rome (1658).
171 Knott, J., (Legends about lice) *Medical Press and Circular* n.s. (1897) **63,** 609.
172 Kolbenm, Peter (*c.* 1745), *Bescreibung des Vorgebürges der Guten Hoffnung,* (trans. Medley) London (1831).
173 Kotzebue, A. F. von (1761–1819), *Erinnerungen aus Paris in Jahre 1804,* Berlin (1804).
174 Krantz, G. W., *Manual of Acarology,* Oregon, 1970.
175 Kurtz, (Phthiriasis) *Magaz. ges. Heilkund.,* (1832) **36,** 97–110.
176 Lancisi, Jerome (1654–1730), *De Noxiis Paludum Effluviis,* Geneva (1718).
177 Landois, L. L. (1837–1902), (Anatomy of the crab louse, body louse and bed bug) *Z.f. wiss Zool.,* (1864) **14,** 1; ibid., (1869) **15,** 32 and 494; ibid., (1868) **18,** 206; ibid., (1869) **19,** 206.
178 Langland, William (1330–1400), *Piers the Ploughman,* (ed. W. W. Skeat) Oxford (1886).
179 Latour, Landry (*c.* 1370), *Book of the Knight,* (trans. Wright) London (1868).
180 Leeuwenhoek, Antony van (1632–1723), *Select Works,* (trans. S. Hoole) London (1707).
181 —, (On the Louse) *Epist. 98,* (trans. Hoole) London (1807).
182 Lehane, B., *The Complete Flea,* London (1969).
183 Lenz, F. W., 'De Pediculo Libellus', *Eranos,* (1956) **53,** 61.
184 —, 'De Pulice Libellus', *Maia,* (n.s.) (1962) **14,** 299.
185 Levinson, H. Z. and Ilan, A. R., 'Assembling and alerting scents

produced by *Cimex lectularius', Experimentia*, (1971) **27**, 102.
186 Lind, James (1716–94), *Two Papers on Fevers and Infection*, London (1763).
187 Linnaeus, C. (1707–78), *Systema Natura*, (10 vols.) Lipsiae (1735–68).
188 —, *Fauna Suecicae*, Leiden (1789).
189 Lorry, A. C. (1726–83), *Tractus de morbis cutaneis*, Paris (1777).
190 Macarthur, W. P., 'Old time typhus in Britain', *Trans. R. Soc. trop. med. Hyg.*, (1927) **20**, 487.
191 Machin, K. E. and Pringle., (Physiology of insect flight muscle) *Proc. R. Soc. (B)*, (1959) **151**, 204.
192 Mackerras, I. M., (ed.) *Insects of Australia*, P. 162, Melbourne (1970).
193 Mackie, F. P., 'The part played by *Pediculus corporis* in the transmission of Relapsing Fever', *Brit. med. J.*, (1907) **ii**, 1706.
194 Madel, W., *Des Flohes Strauss mit der Laus*, Werkztg. Firm C. H. Boehringer No. 3, Ingelheim (1955).
195 Mandeville, Sir J. (? d. 1372), *Travels*, (Cotton MS modernized) London (1923).
196 Marcellus Donatus (1538–1602), *Historia Medica Mirabili*, Venice (1613).
197 Marco Polo (1254–1324), *Travels of Beneditto*, (trans. A. Ricci) London (1931).
198 Marlowe, Christopher (1564–93), *The Tragical History of Dr Faustus*, (ed. H. White) Oxford (1913).
199 Martial (Marcus Valerius Martialis) (43–105), *Epigrams*, (trans. Ker) London (1920).
200 Massaria, Alexander (1510–98). *De Peste*, Venice (1579).
201 Mathiolus, Pietro-Andrea (1500–77), *Comentarii in sex libri ... Dioscorides*, Venice (1554).
202 Maunder, J. W., 'Use of malathion in the treatment of lousy children', *Community Med.*, (1971) **126**, 145.
203 Mayhew, H. (1812–87), *London Labour and the London Poor*, (vol. 3) London (1861).
204 Mayr, E., *Principles of Systematic Zoology*, New York (1969).
205 McKiel, J. A. and West, A. S., (Insect bite reactions) *Pediatric Clinics of N. Amer.*, (1961) **8**, 795.
206 Mead, Richard (1673–1754), (Bonomo's letter to Redi) *Phil. Trans. R. Soc.*, (1703) **23**, 1296.
207 —, *Short Discourses concerning the Pestilential Contageon*, (7th edn.) London (1721).
208 —, *Medical Precepts and Cautions*, London (1751).
209 Meige, H., *Iconographie de la Salpetriere*, Paris (1897).
210 Mellanby, Kenneth, 'The Incidence of Head lice in England', *Medical Officer*, (1941) **65**, 39.
211 —, 'The incidence of scabies in England', *Medical Officer*, (1941) **66**, 141.
212 —, *Scabies*, Oxford (1943).

213 —, 'Symptoms and immunity in scabies', *Parasitol.*, (1944) **35**, 197.
214 —, *Human Guinea pigs*, London (1945).
215 —, 'Reactions to mosquito bites', *Nature*, (1946) **158**, 554.
216 Menagier de Paris (1392), (trans. Power) London (1928).
217 Mercurialis, Jerome (1530–1606), *De Pestilentia*, Venice (1577).
218 Minime, Dr (A. J. Lutard), *La Parnasse Hippocratique*, Paris (1884).
219 Mouffet, Thomas (1553–1604), *Insectorum sive minimorum Animalium Theatrum*, (posth.) London (1634). English version published with E. Topsell's *History of Four-footed Beasts and Serpents*, London (1658).
220 Murchison, C. (1830–79), *A Treatise on the Continued Fevers of Great Britain* London (1862).
221 Nicolle, C. (1866–1936), 'Recherches experimentales sur la Typhus exanthematique', *Ann. Ins. Pasteur*, (1910) **24**, 243. Paris.
222 Nicolle, C., Blaizot, L. and Conseil, E., 'Etiologie de la fievre recurrents', *Ann. Inst. Pasteur, Paris*, (1913) **27**, 204.
223 Noguchi, H. (1876–1928), (*In vitro* culture of spirochaetes) *J. exp. Med.*, (1912) **16**.
224 Nuttall, G. H. F. (1862–1937), Experiments on disease transmission by arthropods, *Johns Hopkins Hosp. Rep.*, (1900) **8**, 1.
225 —, (Comments on Simmond's theory of plague transmission etc.) *J. trop. Med.*, (1902) **5**, 65.
226 —, (Biology of lice and crab lice) *Parasitol.*, (1917) **9**, 293; ibid., (1918) **10**, 80 and 383.
227 Obermeier, O. H. F. (1843–73), (Discovery of spirochaetes of relapsing fever) *Berlin klin. Wschr.* (1873) **10**, 152, 378, 391, 455.
228 Ogata, M., (Concerning Plague in Formosa) *Zbl. Bakt. Orig.*, (1897) **12**, 769.
229 Oudemans, A. C., (The lousy disease) *Zeit. f. Parasitenk.*, (1939) **11**, 145.
230 PAHO/WHO, 'The Control of Lice and Louse-borne diseases', *PAHO Sci. Publ. No. 263*, Washington (1973).
231 Palicka, P. and Merka, V., 'Contemporary epidemiological problems of Scabies', *J. Hyg. Epidem. Microbiol. & Immunol.*, (1971) **15**, 457.
232 Parrish, H. N., 'Deaths from bites and stings in the U.S.A.', *Amer. Med. Ass. Arch. internal Med.*, (1959) **104**, 198.
233 Pasquier, Etienne *et al.* (*c.* 1579), *La Puce de Mme. de Roches*, (ed. Fleuret) Paris (1934).
234 Patterson, B., 'Evolution of vertebrates in relation to parasites', Symposium, *Host Specificity among Parasites of Vertebrates*, Univ. Neuchatel (1957).
235 Paulini, C. F., *Ephemerides naturae curiosum*, Decur. 2 Ann. 5. Append, artic. 38; 42; 60, Leopoldina (1687).
236 Pausanius (Second Century A.D.), *Description of Greece*, (x, x, 6–8) (trans. W. H. Jones) London (1935).
237 Peck, S. A., Wright, W. H. and Gant, J. Q, (Reactions to louse bites) *J. Amer. med. Assn.*, (1943) **123**, 821.
238 Pepys, Samuel (1633–1703), *Diary*, (ed. H. B. Whealey) London (1893).

239 Petrunkevitch, A., 'Spiders', *Encyclopaedia Britannica*, London, etc. (1964).
240 Peus, F., (Fossil fleas in amber) *Palaeont. Zeit.*, (1968) **42**, 62.
241 Phayer, Thomas (1510?–60), *The Regiment of Life, etc.*, London (1545).
242 Pindar, Peter (Pseud. J. Walcot) (1738–1819), *Works*, London (1794–6).
243 Piogey, G., 'Memoire sur la diagnostic de la gale, etc.,' *Gaz. med. Paris*, (1853) **28**, 532.
244 Pitter, Ruth, *Poems*, London (1968).
245 Plato (428–348 B.C.), *The Sophist*, (trans. Fowler) London (1928).
246 Pliny (23–79), *Natural History*, (trans. Rackham) London (1938).
247 Plutarch (*c.* 46–120), *Oeuvres mêlées. Apothegmes des Lacedaemones*, (Sec. iii, vol. 17, p.6) (trans. J. Amyot) Paris (1785).
248 —, *Lives* (iv), (trans. Bernadette Perrin) London (1916).
249 Pollitzer, R., Plague, *Wld Hlth Org. Monogr. No. 22*, Geneva (1954).
250 Pope, Alexander (1688–1744), *Works*, (ed. Elwin and Courthope) London (1871).
251 Power, Henry (1623–68), *Experimental Philosophy*, London (1664).
252 Pringle, Sir John (1707–82), *Observations on the Diseases of the Army in Camp and Garrison*, London (1752).
253 Pringle, J. W. S., *Insect Flight*, Cambridge (1957).
254 Pruss, Johann, *Hortus Sanitatis*, Strassbourg (1491).
255 Purchas, Samuel (1575?–1626), *Hakluytus posthumus or Purchas his Pilgrimmes*, London (1625).
256 Quinte Essence, *The book of* (*c.* 1460), (transcribed Furnivall) London (1866).
257 Rabelais (1483–1553), *Works*, (trans. for Chatto & Windus) London (1886).
258 Radford, C. D., 'Mites parasitic on Vertebrates', *Parasitol.*, (1950) **40**, 366.
259 Rathke, J., (Nature of the 'lousy disease') *Entomologiske Jagttagelser, in Skrifter Naturhistorie Selstrabet*, **5**(1) 192, Copengagen (1799).
260 Raven, C. E., *English Naturalists from Neckam to Ray*, Cambridge (1947).
261 Ray, John (1627–1705), *Historia Insectorum*, London (1710).
262 Reamur, R. A. F. de (1683–1757), *Memoires pour servir a l'histoire des insectes*, Amsterdam (1737–48).
263 Richards, O. W. and Davies, R. G., *Imms Textbook of Entomology*, London (1957).
264 Redi, Francesco (1621–97), *Opusculorum: pars prima sive Experimenta circa Generationem Insectorum*, Amsterdam (1686).
265 —, *Esperienze intorno alla Generazione degl'Insetti*, Florence, 1668 (trans. M. Bigelow), Chicago (1909).
266 Riek, E. F., (Lower Cretaceous Fleas) *Nature*, (1970) **227**, 746.
267 Rihab, Mohammed, 'Der Arabische Artz al Tabari', *Arch. f. Ges. Med.*, (1927) **19**, 123.
268 Robathan, Dorothy N., *Ovid in the Middle Ages*, (Ch. 6 of '*Ovid*', ed. Binns) London and Boston (1973),
269 Rosenhof, Rosel von (1705–59), *Insecten Belustigung*, ii, 'Sammlung, der

Mücken and Schnaken' Nürnberg (1749).
270 Rothschild, Miriam and Ford, R., (Pheromones and flea reproduction) *Proc. R. ent. Soc. Lond. A*, (1970) **35**, 1.
271 Rothschild, Miriam and Hinton, H. E., 'Holding organs of antennae of male fleas', *Proc. R. ent. Soc. Lond. A*, (1968) **43**, 105.
272 Russell, Bertrand, *History of Western Philosophy*, London (1946).
273 Russell, John (*c.* 1450), *The Boke of Nurture*, (ed. Furnivall) London (1866).
274 Salberg, J. J., (Experiments on insecticides for use against bed bugs), *König Schweden Acad.*, Abh. 7, p. 20, Stockholm (1745).
275 Salmon, W. (1644–1713), *New London Dispensary*, London (1676).
276 Sauvage de la Croix, F. B. (1706–67), *Nosologica Methodica*, Lyons (1759).
277 Scaliger, Julius Caesar (1484–1558), *Exercitationes de Subtilitate ad Hironymum Cardanum*, Lutetiae, Paris (1557).
278 Schierbeek, A., *Jan Swammerdam (Life and Works)*, Amsterdam (1967).
279 Schjödte, J. C., (Discussion of phthiriasis and mouthparts of the louse) (trans. from Danish) *Ann. & Mag. Nat. Hist. Lond.*, (1866) **17**, 213.
280 Schrut, A. H. and Waldron, W. G., (Entomophobia) *J. Amer. med. Assn.*, (1963) **186**, 429.
281 Serenus, Samonicus, Q. (d. 1212), *De Medicina Praecepta*, (ed. I. C. G. Ackerman) Leipzig (1786).
282 Sergent, E. and Foley, H., 'Recherches sur la fievre recurrente', *Ann. Inst. Pasteur, Paris*, (1910) **24**, 337.
283 Shakespeare, W. (1564–1616), *Concordance*, (ed. John Bartlett) London (1927).
284 Sharov, A. G., (Basic arthropodan stock) Oxford (1966).
285 Shrank, A. B. and Alexander, S. L., 'Scabies: another epidemic?', *Brit. med. J.*, (1967) **i**, 669.
286 Sikora, Hilda, (Biology of lice) *Zbl. Bakt. 1 Abt. Orig.*, (1915) **76**, 523; *Arch. Schiffs. & tropen Hyg.*, (1916) **20**, Beiheft 1, 1.
287 Simond, P. L., 'La Propagation de la Peste', *Ann. Inst. Pasteur*, (1898) **12**, 625.
288 Singer, Charles, *A Short History of Scientific Ideas*, Oxford (1959).
289 Singer, CK. AND Singer, D., 'History of discovery of microbes', *17th Internat. Cong. Med. (Hist. Med.)*, London (1914).
290 Snodgrass, R. E., (Anatomy of Fleas) *Smithsonian Miscl. Coll.*, **104**, Washington (1946).
291 Southall, John, *A Treatise on Buggs*, London (1730).
292 Southey, Robert (1774–1843), *Southey's Common Place Book*, (ed. Warter) London (1849–51).
293 Sperling, Johann (1603–58), *Zoologia Physica*, (Posth.) Lipsice (1661).
294 Steadman, A., *Wanderings and Adventures in the Interior of S. Africa*, London (1835).
295 Steinhaus, E. A., *Insect Pathology* (2 vols.) New York and London (1963).

296 Swammerdan, J. (1636–80), *The Book of Nature*, (trans. T. Flloyd, ed J. Hill) London (1758).
297 Swift, Jonathan (1667–1745), *Poems*, (ed. H. Williams) Oxford (1958).
298 Talmud, (Second to sixth century A.D.) (trans. Epstein) London (1938).
299 Tictin, J., (Transmission of relapsing fever: bugs suspected) *Zentbl. Bakt.*, **21**, (1897).
300 Tiegs, O. W. and Manton, S. M., (Evolution of Arthropods) *Biological Reviews*, (1959) **33**, 255.
301 Tissandier, G., *Recréations Scientifiques*, Paris (1876).
302 Titschack, E., (Biology of bed bug) *Zeit. Morph. Oekol Tiere.*, (1930) **17**, 471–551.
303 Traub, R., 'Convergent evolution in Fleas', *Bull. Brit. Mus. (Nat. Hist.) Zoology*, (1971) **22**, (12).
304 Tryon, Thomas (1634–1703), *A Treatise on Cleanness in Meats, etc.*, London (1682).
305 Turner, David (1667–1741), *De morbis cutaneis*, London (1714).
306 Turner, E. S., *The Court of St James*, London (1959).
307 Tusser, Thomas (1524?–80), *Hundreth Good Points of Husbandrie*, London (1557).
308 Usinger, R. L. *et al.*, *Monograph of Cimicidae*, Thomas Say Foundation VII Ent. Soc. Amer. (1966).
309 Usinger, R. L. and Povolny, D., (Original population of bed bugs) *Acta Musei Moraviae* (1966) **51**, 237.
310 Valentin, M. B., (Case of lousy disease) *Acta phys. med. Acad. Caes. Leop. germ. Nat. Cur.*, (1730) vol. 2, p. 396.
311 Vallisnieri, Antonio (1661–1730), *Esperienze ed Osservazione intorno all' Origini, Sviluppi e Costumi di varij Insetti*, Padua (1726).
312 —, (Case of lousy disease) *Opere physico-mediche*, etc. vol. 1, 337; *Nov. giunta*, pp. 339, etc., Venice (1733).
313 Wafer, Lionel (?1600–1705), *A New Voyage and Description of the Isthmus of America*, (in Capt. W. Dampier's *Voyages Round the World*) London (1729).
314 Walckenaer, C. A. (1771–1852), *Histoire naturelle des Apteres*, Paris (1844).
315 Wallace, A. R., (Louse eaters of S. America) *Trans. ent. Soc. Lond.*, (1852) **2**, 241.
316 Wanley, Nathaniel (1634–80). *The Wonders of the Little World*, (Bk. 6, Ch. xx) London (1678).
317 Weidner, H., '*Die Flöhe* (Siphonaptera) Unterfrankens', *Beitr. z. Insektenfauna Unterfranken mit Naturwiss Museum No. 13.*, Aschaffenburg (1973).
318 Whitehead, G. B., (Insecticide resistance of bed bugs in S. Africa) *J. ent. Soc. S. Africa.*, (1962) **25**, 121.
319 Wichmann, J. E. (1740–1802), *Aetiologie der Krätze*, Hanover (1764).
320 Wigglesworth, V. B., *Insect Physiology*, (6th edn.) London (1965).
321 Willan, Robert (1737–1812), *On Cutaneous Diseases*, London (1808).
322 Willart De Grecourt (1683–1743), *Oeuvres Divers*, Luxembourg (1761).

323 Wither, G. (1588–1667), *Britain's Remembrancer: containing a Narrative of the plague lately Past,* (1628) Manchester (1880).
324 Wolfart, K. C. A. (1778–1832), (Spontaneous generation of lice) *Aekahmieon,* (1818) No. 45, 705.
325 Woodforde, J. (1758–1802), *Diary of a Country Parson* (ed. Beresford) Oxford (1949).
326 Wotton, Edward (1492–1555), *De Differentiis Animalium,* Paris (1552).
327 W. W., *The Vermin-Killer, being a necessary Family book, etc,* London (2nd edn. 1680).
328 Yersin, A. and Roux, E., (Plague and rats) *Bull. Acad. Med. Paris,* (1897) **37,** 91.
329 Zdrodovskii, P. F. and Gdinerich, H. M., (The Rickettsial Diseases) (trans. B. Haig) Oxford (1960).
330 Zinsser, H., *Rats, Lice and History,* (4th edn.) London (1942).
331 Ziegler, P., *The Black Death,* London (1969).
332 Zwinger, Th. (1533–88), *Theatrum humanae,* Lib. iv, 373; vii, 525, Basel (1571).

Index of Persons

(Ancient authors have their life dates appended. Other names are of 20th-century authorities)

Subject Index